Writer's Handbook

FOR ENGINEERING TECHNICIANS AND TECHNOLOGISTS

DAVID W. RIGBY

North Seattle Community College

The Wordworks™ Series

Prentice Hall

Upper Saddle River, New Jersey
Columbus, Ohio

Library of Congress Cataloging-in-Publication Data

Rigby, David W.
    The writer's handbook for engineering technicians and technologists / David W. Rigby.
        p. cm. -- (Wordworks)
    Includes bibliographical references and indexes.
    ISBN 0-13-490103-7
        1. English language--Technical English--Handbooks, manuals, etc. 2. English
    language--Rhetoric--Handbooks, manuals, etc. 3. English
    language--Grammar--Handbooks, manuals, etc. 4. Technical writing--Handbooks,
    manuals, etc. I. Title.

PE1475 .R56 2001
808'.0666--dc21

00-034707

Vice President and Publisher: Dave Garza
Editor in Chief: Stephen Helba
Executive Editor: Debbie Yarnell
Associate Editor: Michelle Churma
Production Editor: Louise N. Sette
Production Supervision: Clarinda Publication Services
Design Coordinator: Robin G. Chukes
Text Designer: Ceri Fitzgerald
Cover Designer: Ceri Fitzgerald
Production Manager: Brian Fox
Marketing Manager: Jimmy Stephens

This book was set in Optima by The Clarinda Company. It was printed and bound by Victor Graphics, Inc.
The cover was printed by Victor Graphics, Inc.

The Wordworks trademark is the registered property of David W. Rigby. © 2000 by David W. Rigby.

10 9 8 7 6 5 4 3 2
ISBN 0-13-490103-7

Welcome to Wordworks™

Wordworks™ is a series of four communication skills manuals. The manuals consist of three writers' guides for engineering and technical applications and an additional guide to in-service spoken communication. The manuals are designed to provide in-demand information in a readable fashion. They are matter-of-fact and use-oriented.

Each manual focuses on specific exit skills that are necessary for job performance. In this respect the texts are somewhat unique. They were inspired by the carefully tailored goal orientation of corporate seminar manuals. This strategy is at the heart of the streamlined manuals of corporations where specific skill outcomes are the narrow focus of company class time. For skills manuals this strategy can accelerate and focus the learning process, and the approach is particularly useful in college programs where English components are never more than a course or two of the total learning experience.

The *Wordworks™* manuals are conversational, visual, and practical so that the learning experience is accessible. Chapter-by-chapter discussions encourage a learning curve of understanding. Important concepts and practical applications are identified and explored. Models and other illustrations are also features of the texts.

The texts rely on an extensive use of graphics to conceptualize ideas, and models are provided to draw attention to desirable skills. This approach is intended to help students build a strong understanding of logical design features that they can use to construct any writing project.

Three of the texts in the series deal with the basics of the craft of writing, and the series uses a writer-to-writer strategy to explore and explain this craft. The manuals are intended to be learning tools, but because they focus on a craft, they are not intended to be overly academic. To the extent to which streamlining can be achieved in a college environment, the *Wordworks™* series provides a practical approach to skills training that is compatible with the limited time available for communications offerings in technical programs. The manuals encourage curiosity and provide a learning-oriented climate, but they also simplify the path to practical knowledge and skills.

Each title in the *Wordworks™* series is intended to complement the other titles:

Basic Composition Skills for Engineering Technicians and Technologists. This first text of the series is intended to help upgrade fundamental skills in writing. It is a thorough discussion of the problems that are encountered by writers and the solutions to those problems. This book is uniquely designed to build upon existing skills that are part of every work day.

Writer's Handbook for Engineering Technicians and Technologists. The second title in the series is a writer's handbook of the rules and practices of writing. Part style guide, part grammar book, part technical writing reference, the *Writer's Handbook* is designed to be a bridge that can support both *Basic Composition Skills* and *Technical Document Basics*. The *Writer's Handbook* is specifically intended for engineering and technical students.

Technical Document Basics for Engineering Technicians and Technologists. The third title in the of *Wordworks™* series develops a concentrated focus on basic technical writing know-how. The text is designed to identify and explore the documentation standards that are used to develop and produce technical projects. The basic skills are condensed into a short, readable text.

Workplace Communications for Engineering Technicians and Technologists. An additional title supports the *Wordworks™* concept with an exploration of spoken communication. Studies reveal that 80% or more of our work-related communcation in trade and industrial settings is handled in conversation. The absence of training tools in this area is an invitation for communication problems if only because spoken messages outnumber our memos by four to one! *Workplace Communications* helps identify and improve in-service communication skills.

About the Writer's Handbook

Engineers and engineering technicians are not simply "numbers people." We write a great deal, and usually our incomes depend partly or entirely on our writing—of contract proposals, for example. We need to be able to handle the language and purge our errors before the department supervisor or our clients see our documentation.

A host of dilemmas, including some that are unique to engineering and technical documentation, can emerge when we write. Do we put Latin technical words in italics? Do we capitalize *PCB* if we do not capitalize the term *printed circuit board*? Do we write with the "I" pronoun? Was this *phenomenon* an *affect* or an *effect*? Or was the correct word *phenomena*? In addition, there is the vast vocabulary of any technical speciality—biomedical electronics, for example—that cannot be found in the dictionary:

apneustic flowtime

hemodynamic pressure

vectorcardiography

electroencephalography

sphygmomanometers

The spell checker will identify many errors in spelling, including typos. However, it will not correct either technical vocabulary or such writing errors as the notorious "rhyme pairs" whereby you spell one word but mean another (for example "cite" becomes "site".)

Because engineering, technological, and scientific writing involves very difficult material, the basic task we face in writing is the challenge of clear communication. If writing is organized and controlled with enough skill, a reader can navigate through almost any document. Good communication depends on clarity, and clarity depends on understandable writing. For the technical content of our work, clarity may be almost impossible to achieve at times, so it is important to realize that the vehicle—the language—can be crafted for clarity. Writing that creates a path of least resistance for the reader is our goal. Any incorrect calculations, any grammar errors, or any vague concepts will interrupt the reader's comprehension of the document.

The *Writer's Handbook* is a desktop guide that is designed to address the specific skill needs for writing projects in college and at work. It consists of hundreds of rules that can help eliminate the flubs and bobbles that find their way onto the computer screen. The rules are organized so that writers can consult the various sections of the book as needed. Readers should initially read the text from cover to cover to become familiar with both its organization and the many, many controls that are explored in the manual. The text is written in a manner that should help make the rules vivid and logical.

The *Writer's Handbook* consists of chapters that deal with sentence logic, punctuation logic, grammar logic, logic lapses, technical supports (numbers, measures, abbreviations, symbols), math displays, and source notations. The text begins with the basics, a discussion of the design of logical sentences. This section is followed by an extensive presentation on punctuation, which is an important aspect of sentence logic. The fundamentals of grammar are presented in the third chapter; this material is also a critical element in sentence construction. Another chapter concerns spelling and capitalization.

There is a chapter dedicated to the common errors and the frequent problems that occur in engineering documentation and technical writing. This slip-and-oversight category is rather large and accounts for a host of common errors.

Although all the material in the text is tailored for engineering tech students and engineering students, the uses of mathematics, measures, abbreviations, and symbols are a unique concern in scientific and technical areas. A related chapter that is designed to explore the unique world of "writing" mathematics follows the discussion of measures and abbreviations. A subsequent chapter discusses the practical matters of citing and managing resources, since the reference list, in particular, is a very important feature of many technical documents. The final chapter of the *Writer's Handbook* is dedicated to resume writing.

When there is little need to explain a rule, the rules are bulleted. Many rules simply say "do this" or "do not do that." However, other rules call for explanations. Also, at times the text explains elements that are not really rules at all, such as the word families we call "parts of speech." These elements of writing are the accepted conventions of the language. Engineers often refer to "controls," the engineering parameters by which they must abide. In a sense, all the rules and standard practices of our language are controls—and the *Writers Handbook* is a book of controls.

Writing is one-third thinking and one-third production management, which are the concerns of *Basic Composition Skills* and *Technical Document Basics,* the two titles of the *Wordworks*™ series that explore writing strategies in engineering and engineering technologies. The final third consists of the fine points—the detailing. The *Writer's Handbook* contains all the essentials a writer needs to know in order to control the quality of the fine points. Remember, the quality of a document is in the hands of the writer, and the writer should be familiar with at least one writer's guide when there are questions about the "rules." The *Writer's Handbook* is here to help.

Contents

Chapter 3

Sentence Symbols: Punctuation Logic   51

Chapter 4

Sentence Conventions: The Basics of Grammar   93

Chapter 5
Word Rules: Spelling and Capitalization   133

Chapter 6
Goofs, Gaffes, Glitches, and Gremlins   159

Chapter 7

Numbers, Measures, Abbreviations, and Symbols   223

Chapter 8

Symbolic Systems: The Languages of Science   255

Chapter 9

Cues and Keys: Citations and References   277

Chapter 10

Internet Resources for Engineering   317

Chapter 11

The Resume: Your Corporate Passport   357

Appendix A

Getting Around in the *Writer's Handbook*

Basic *Composition Skills* explored the fundamental patterns and practices for paragraph logic and the essentials of document design. *Technical Document Basics,* the third volume in the *Word-works* series, also examines these large components, and that text explains specific applications in document architecture for technical projects. The *Writer's Handbook* is designed to focus attention on the smaller features of language construction—though they are no less important in the scheme of things.

The design of the the *Writer's Handbook* is based on a simple organization principle that you should recognize as a familiar pattern: scale. Look at writing in terms of units that decrease in size from large to small. Atomic structure demonstrates one such pattern:

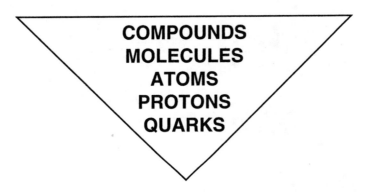

COMPOUNDS
MOLECULES
ATOMS
PROTONS
QUARKS

The language system is similar; it, too, has greater and lesser component parts:

DOCUMENTS
PARAGRAPHS
SENTENCES
PHRASES
WORDS

In the logic of the language triangle, lesser units are grouped into the components that make up the next unit, and each unit is a composite made up of the smaller units: words make up clauses, clauses make up sentences, sentences make up paragraphs, paragraphs make up documents. You could even take the simplest unit—words—and break that level down into the alphabetic letters.

The hierarchy in the triangle of language components is a handy way to divide the rules into categories. Just as there are universal principles controlling atomic structure, there are standard practices—the controls or rules—that regulate the systems that operate at each level of the language.

RULES FOR DOCUMENT DESIGN
RULES FOR PARAGRAPH LOGIC
RULES FOR SENTENCE LOGIC
RULES FOR SENTENCE SYMBOLS
RULES FOR SENTENCE CONVENTIONS
RULES FOR CLAUSES
RULES FOR WORDS

This text focuses on the bottom five levels of this structure. It examines the correct detailing of sentences, phrases, and words, in terms of grammar, mechanics, punctuation, spelling, and other conventions.

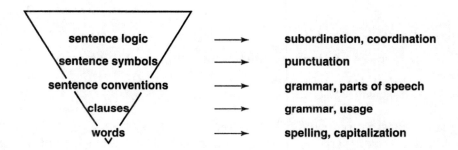

sentence logic	→	subordination, coordination
sentence symbols	→	punctuation
sentence conventions	→	grammar, parts of speech
clauses	→	grammar, usage
words	→	spelling, capitalization

Each element of the triangle is developed into a chapter in the *Writer's Handbook*.

The sentence is the basic unit of the logic system of language. If the sentence does not function, the entire thought system collapses. You can look at sentences from several perspectives. First, there are controls for the *logical development* of ideas, and these affect the construction of sentences. A system of demarcations—*punctuation*— holds the pieces of the language together. Somewhat different controls are the grammatical *conventions* used so that people can "share" the language. These three elements of the sentence are considered in separate chapters.

One chapter examines the principles of coordination and subordination and the logical processes of causation and parallelism, which are part of the logic structure of the language. Logic is one tool writers use to design sentences, so it is helpful to see language as a system of logical structures.

A second tool for sentence development is punctuation, which also merits a chapter. The elements of punctuation are unique in being silent partners (they are not spoken) and in being "function signs" or symbols that execute a logic process of some kind. (They can be quite mathematical, as you will see.) Punctuation symbols are familiar to everyone, of course, but writers puzzle over the applications of some of them, particularly the semicolon and colon. Both of these devices are critical assists in technical documentation. In general, all the symbols are convenient logic links that bind the logic of sentences. About half of the symbols are used internally to identify and relate parts of a sentence. The other half are used to mark the parameters of sentences or to link them into larger units. All the devices support the sentence logic writers are attempting to structure as they write.

A subsequent chapter begins to shift your attention to grammar rules because there are traditional methods that are used to construct everything from verb conjugations to pronoun agreements. These practices are highly standardized, and misuse of the rules is often noticeable. Verb errors are usually obvious:

> You got to get the job done.

Any pronoun agreement problem is simply a logical inaccuracy:

> The student must register first, and then they pay the tuition fee.

The discussion of sentence conventions includes a review of parts of speech. This chapter concludes with a look at sentences and sentence parts from the viewpoint of grammatical conventions.

Several chapters look at controls or rules for the use of words and the systems for properly using these small units of language. The rules and regulations may seem endless, but there are underlying reasons for the laws that you will find important. Consider spelling, for example. Writers *do* need to spell correctly—but not because they will be misunderstood. That is what you are always told, and the idea is simply not true. I perfectly well understand the *worst* speller. The problem is that the author looks foolish. In other words, there are *logical* and *social* considerations to examine among the details of even the smallest units of the language. Writers want to appear logical, and they do not want to appear to be wrong. You need to look at spelling practices and spell-check applications. You also want to be on the lookout for dedicated software that might identify the vocabulary, the symbols, and the abbreviations that are applicable to your field.

The longest chapter, as you might guess, concerns all the usual errors that find their way into writing. Although the other chapters tend to discuss what to do to produce good writing, Chapter 6 takes the opposite approach and points out what *not* to do. The chapter identifies most of the problems that technicians and engineers seem to grapple with in the course of their written work and that result in a curious mix of very technical and very simple errors. These problem areas include pronoun confusion and verb errors, and sentence component problems with modifiers, for example. Included in this chapter are lists of common misuses of a wide variety of popular expressions.

Several chapters are included on technical matters that do not usually appear in writers' guides. These sections involve math systems, abbreviations, and other technical considerations of interest to technicians and engineers. The subsequent chapters discuss the citation and reference methods used in research, and survey engineering resources and the Internet. The final chapter concerns resume writing.

This manual does not uniformly present "rules" in the sense of unexplained regulations. Whenever practices are obvious, I simply state controls (rules) or present samples of conventions. Many grammar books use this approach. An equal number of manuals explore the writer's world at length and provide in-depth discussion of the *practices* of the writer's craft. As necessary, I include discussions of practices among the rules and regulations. When the issues are likely to need clarification, I explore them in some detail. If any point of discussion remains hazy, you would do well to consult a more detailed grammar.

The rules selected for the *Writer's Handbook* were chosen for engineering college students, particularly at the two-year-program level, and the material had to be selected on the basis of skill levels and skill emphases for these writers. Depending on your particular skill needs, you may want to seek ESL materials, basic grammar books, or the more thorough college transfer grammar books located in college bookstores.

Cybergrammar

At the outset, you should look at the benefits and limitations of software applications designed for writers. Language is much too flexible and variable for computers to edit thoroughly. The systems cannot judge the correctness or the logic of your work. Sentences are highly organized devices, but they can be rendered in a wide variety of patterns. If I were to calculate the mathematical permutations of combinations of twenty words in sentences and a vocabulary of, say, ten thousand words, the results would be in the millions. Or to use statistical probabilities, if a typical public library holds one hundred thousand books, what do you suppose is the likelihood that the same sentence appears in any two of the books? The probability is near zero. The variables are hardly programmable. Fully dependable grammarware is not a practical option in the near future.

At present grammar checks are reasonably effective on matters concerning passive verbs, redundancies, jargon, slang, sexist terms, contractions, and punctuation in quotations. I

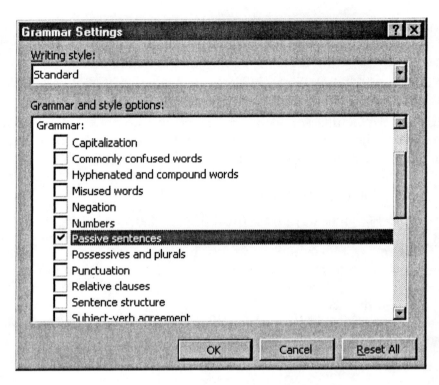

A few of the grammar checker options (Screen shot reprinted by permission from Microsoft Corporation.)

located several dozen passive constructions using this book as a test. At the other end of the spectrum, the programs are of little value in checking coordination, subordination, parallelism, point-of-view shifts, and sentence variety.

Although grammar checkers will flag apostrophe errors, and some semicolon errors, the programs are of little help with commas, colons, and the other punctuation marks, or with underlining, italics, and boldface. The grammar check suggestions present mixed results in many areas:

verb errors

subject-verb agreement errors

missing verbs

sentence fragments

missing or misused articles

gerunds and infinitives (after the verb)*

* *These comments represent an approximate tally of the findings of Diana Hacker, reported in* **The Bedford Handbook** *(Boston: Bedford Books, 1998).*

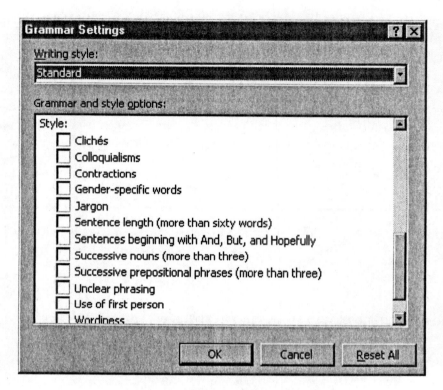

A few of the grammar checker style options (Screen shot reprinted by permission from Microsoft Corporation.)

The success rate may improve as programs improve, but the language flexibility remains the challenge. For example, although the rules of capitalization are cut and dried, a computer cannot make a judgment call on a proper noun because this is a social decision.

You can use the grammar checker selectively since the popular Microsoft versions have checklists that allow you to tailor the editing suggestions. Otherwise, the help balloons will just keep coming. Because the checker flags possible errors, you can waste considerable time checking numerous possible problems that are false alarms. Select the high-risk areas where you want the computer to identify potential trouble, though do not expect "corrections." You may find glaring errors or only revision suggestions, but either way the help is useful.

The spell checker, on the other hand, is a wonderful tool of precision. The results are precise and immediate. It is comforting to know that there are technical assists that can improve our writing—and at remarkable speeds. Besides, a great many errors are typos, and the spell checker is particularly efficient in sorting out such problems because typos usually are either in the words of the text or between words, and in both cases the program reads the errors instantly.

The existing grammar programs can "calculate" interesting and possibly useful information that functions more as style observations than as corrections. The typical grammar program takes a statistical focus. It will tell you, for example, the number of paragraphs you have created, the number of sentences, the number of words, the number of questions, and so on, including the average length of your words and sentences. Unfortunately, these measures are not much of an "edit," although I did find out that each volume of *Wordworks* involves about three million characters or bytes (one byte equals one letter). It is little wonder that errors are unavoidable! The following sample printout is typical of these software applications:

Problems detected: 17	Applied Comp	Fri Oct 25 16:55
DOCUMENT STATISTICS	INTERPRETATION	
Grade Level: 10	Preferred level for most readers.	
Reading ease score: 51	Average reading level. 6–10th grade level.	
Passive voice: 19%	Writing may be difficult to read or ambiguous for the chosen writing style.	
Avg. sentence length: 13.8 words	Choppiness or overuse of short sentences may be indicated. Try varying sentence lengths.	
Avg. word length: 1.67 syl.	Most readers could comprehend the vocabulary used in this document.	
Avg. paragraph length: 8.4 sent.	Most readers could easily follow paragraphs of this length.	

Another particular program determines the number of passive constructions the author created, and this function has editing and stylistic value. The program also uses a reading-skill analysis to predict how well the target audience of the document will understand the document. If the criteria of the reading analysis are reasonably accurate and valid, the reading skill level is a useful consideration for documents of a technical or scientific nature. This sample program provided the author with the following suggestions:

```
Statistics for Applied Composition Project 3
Problems marked/deleted:  0/17

Statistics:      Readability
                 Reading Ease:              51
                 Fog Index:                 13
                 Grade Level:               12

Statistics:      Paragraphs
                 Number of paragraphs:      14
                 Average length:             8.4 sentences

Statistics:      Sentences
                 Number of sentences:       82
                 Average length:            13.8 words
                 End with "?":              0
                 End with "!":              0
                 Passive voice:             8
                 Short (< 14 words):        57
                 Long (> 30 words):         2

Statistics:      Words
                 Number of words:           1083
                 Prepositions:              60
                 Average length:             5.07 letters
                 Syllables per word:         1.67
Count categories:                           NONE
```

The problem with an analysis such as this one is that it will overlook logic flaws such as the following:

> A fatal situation such as death may occur.

The sentence says that death will result in death. The computer did not and could not detect this problem. There is no substitute for close scrutiny of the logic of writing and a manual of rules that can be used by the writer. As noted earlier, the plasticity of our language allows for too many variants. The variables cannot be programmed because the rules of writing are *contingent* rules that say, "Do A but . . ." or "Do B except . . ." or "Do C if . . . ." The "if rules" are the frustration of our computers. They are a frustration any author knows all too well. Writing takes more than electrons. Writing takes neurons. Any misapplication of common words will easily slip by the scrutiny of a machine. Look at these comical errors that went straight from the keyboard to the printer without a hitch:

Ten pounds was not much of a wait loss.

She enjoyed matched potatoes.

Urban violence, crime and unemployment are all the results of the replacement of labor by technology such as computers and rowboats.

The heavy steel pattern is then lifted off by hand or with wenches.

School suctioned prayer should be banned.

It was illeagle.

I think in pitchers.

School is the hardest thing that I have been struck with.

Every one of these zingers from English remedial classes managed to slide by computer checks quite smoothly. In fact, I suspect that several of these errors came from helpful suggestions from friendly computers that offer "similar-word" search features when the spelling of a word is not clear. Since no definitions are provided, the authors, particularly ESL authors, can become quite confused. There is some truth to that popular sign that hangs in the computer labs all over the country:

> **To Err is Human.**
> **To Really Foul Things up**
> **Requires a Computer.**

If you use a handheld caculator or mathematics applications in your computer, you can appreciate the amazing precision of these machines. There are no two ways about it when you deal with math. Language is quite the opposite. No software application will carpet-bomb your writing projects. Basically, all the controls your computer can access for you are quantity controls. Quality controls do not program in hexadecimals. Quality is up to you.

When you leave your favorite software store without having been able to find good supportware for your term paper or your lab reports, you realize that writing is a craft that plays by the rules but remains flexible and creative. Grammar check features are reliable, but only to a point. When you are writing, your computer is functioning as a word processor, and it is depending on you. You need to know the rules of the game and how to apply those rules to your writing.

Playing by the Rules

"A vital few of the causes account for most of the total defect, while the trivial may account for very little of the total defect."

Vilfredo Pareto (1848–1923)

What exactly are you supposed to do with a book of rules? Should you try to apply all the rules all the time? Should you try to study the rules you do not know? Should you consult the rules when you are in doubt? Or should you run to the rules once your grade plummets or your boss has torpedoed a recent set of badly handled reports?

Few books of rules and regulations, regardless of the field or topic, are going to be anybody's preferred reading. The rules come thick and fast, and the reader prays for osmosis. In truth, if you use the rules, you will tend to remember them; you may have difficulties with the rest. The question then remains, Is there a tactic that might improve the return on your use of a rule book?

You will seldom see any books of regulations or of tables that have a plan of action. Grammar books usually offer exercises, but it is not possible to individualize them. You will need a strategy to address your specific needs.

One tactic is to look at your writing in terms of defects. The rule books tell readers what to do correctly, but this information does not address your needs in particular. Look at your errors with the eyes of a statistician. Statisticians are fond of a popular theory proposed by the Italian social theorist Vilfredo Pareto, who simply observed that three-fourths of a group of defects are caused by one-fourth of the possible causes. Thus, it is likely that if you remove a few of the major causes of writing errors, you will remove a great many writing defects.

There is no need to understand the complications of Pareto Charts and other statistical implications of the theory, but you do need to look at the technique for analyzing defects and their causes. Since you are looking at editing errors in a document, what you want to know is the *frequency* of an error and its *importance*. In other words, you need to rank your errors. The frequency of a single error may be important, or a single error of another type may be equally significant.

Any high incidence of error—perhaps in verb use or pronoun use—is an area of focus. For example, there are a large number of rules for the proper use of numbers. It is very likely that you use numbers extensively in your writing, and it is also likely that you have seen few if any of the numbers rules. Thus, this will be a predictable high-error area, but if you learn two dozen rules, you can render hundreds of numbers correctly in a single document. Similarly, if past tense verb endings leave a trail of twenty errors in a two-page business letter, mastering one tense will solve the problem. If you strengthen the weakness, you overcome the problem.

Prioritize errors.

Note the frequency of errors.

Identify the maximum return for your investment and focus on errors you can address in volume. Use a rule book to identify the *few* possible causes of many of your problems, and you will gain confidence in your writing.

The Code Key: Editing Your Work

Any writing project in the initial stages is a magnet for errors. Although there is no reason to be careless in the rough-draft stage of development, there is little reason to groom the document before it is fully composed. To correct your drafts you will most likely proofread on the computer screen. At any point you can use the spell checker to quickly correct spelling and spacing errors. The grammar checker may best serve you after the document is in an advanced draft, at which point the document is ready for precise editing. You should find the following code helpful once you generate hardcopy. Simply use a colored pen and correct the text. The editing code is handy because it allows you to indicate changes quickly without taking the time to either develop or investigate the actual corrections that are needed. The code is also valuable as a communication tool if you have the opportunity to have someone else edit the production. It is at this final stage of development when the time-saving and space-saving utility of the code is of most value.

The following six editing symbols appear with great frequency in proofreading edits. They usually are used in conjunction with editing abbreviations that further explain the suggested revision.

Proofreader's Marks

X	error or typo
^	insert
~	transpose
⌒	close the space
℘	delete
/	Delete or separate

Any of the following abbreviations may be used. Expect variations on these; in addition, editors and college instructors develop other personal markings that will be helpful to you. For example, I simply tend to circle errors.

The Proofreading Code

abb	abbreviation	**log**	logic
acc	check accuracy	**mm**	misplaced modifier
ad	adjective/adverb	**ms**	manuscript form
agr	agreement	**num**	number
awk	awkward	**ok**	let stand
bw	better word available	**¶**	paragraph
cap	capitalization	**no ¶**	no paragraph
co	coordination	**//**	faulty parallelism
coh	coherence	**para**	paraphrase
conn	connotation slants	**pass**	passive construction
conc	conclusion	**ref**	unclear pronoun reference
cs	comma splice	**run-on**	run-on or fused sentence
def	define	**shift**	shift
dev	develop text	**slang**	slang
dm	dangling modifier	**slant**	language bias
doc	documentation needed	**sp**	spelling
ex?	example needed	**sub**	subordination
frag	sentence fragment	**sum**	summarize
fs	fused sentence	**tr**	transpose
id	idiom	**trans**	transition
inc	incomplete construction	**vague**	vague
intro	introduction	**verb**	verb form
ital	italics	**wrdy**	wordy
jarg	jargon	**ww**	wrong word
lc	lowercase (no cap)	**ww?**	check word use

Punctuation corrections usually involve either removal or insertion of punctuation, which is indicated by a slash or a caret, respectively.

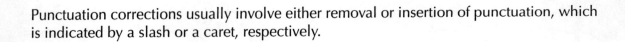

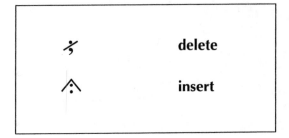

	delete
	insert

The slash and the caret can appear with any of the punctuation marks:

. ? !	**period, question mark, exclamation point**
,	**comma**
; :	**semicolon, colon**
'	**apostrophe**
italics **Bold**	**italics, bold**
_____	**underline**
( ) [ ] –	**parentheses, brackets, dash**
" " . . .	**quotation marks, ellipses**

Running Texts and Displays

There is one term that appears frequently in the *Writer's Handbook* that is not identified in the preceding code. Editors use the expression *running text* to mean the sentences in the paragraphs that represent the body of the document. This handbook will frequently use this expression to discuss "text" practices, which in technical and engineering documentation are quite distinct from "display" practices. Displays refer to material that is set off from the text, typically including mathematics, chemistry, physics, and most engineering calculations.

Exercises Chapter 1

1. Read the following rough draft and identify errors with the appropriate editorial markings.

2. Rewrite the document and make the appropriate corrections.

Roadbuilding is an anceint craft and many major civilizations depended on roads to develop and defend their lands. The Romans were excellent road builders and there are roads in modern Italy that rest on the ancent beds of early Roman roads stretches of which can still be seen today.

The Empire of the Incas depended on their roadway system which traversed much of western South America. Because of the Andes Mountain Chain most of the roads ran north and south on the Pacific side of the Andes although there were roads that penetrated the Amazon basin. It was because of the roads that a smallband of Spanish explorers found the Incan capitol city of Cuzco Since South America was uncharted wilderness there was little chance that the Spaniards would have survived long enough to find the Incan capital otherwise.

North Americans faced the chalenge of western expansion by using Engineering skills to explore and expand the national horizon. In the 1820s new technologeis were put to work to build canals. The Erie Canal took eight years to build and traversed 363miles from Lake Erie to the Hudson River. At about the same time the first important road was complete. The National Road ran from the shores of the Potomac to Wheeling West Virginia. A third entry among new transportation technologies of North Amercan was the Locomotive. Between 1830 and 1890 2500 miles of track each year were put down by the Amercans. In the year 1869 the transcontinental railroad marked the end of the Western expansion. A new emphasis—speed—was soon to change the nation. Barges on the Erie Canal moved slowly. Trains averaged 25 miles per hour and even as late as 1938 the journey from Chicago to New York on the New York Central took 16 hours. However the popularity of the automobile forever changed the national perception of travel. In the 1920s the American Government established the Public Roads Administration, and by the 1930s the new highways were intersected by the new cloverleaf intersections at regular intervals which *could be seen from thee passenger seat of the Boeing 247 the countrys first twin engine passenger plane (with 10 passengers) to enter commercial service. The year was nineteen thirty three!*

3. If you have Microsoft Word 97 or Word 2000 on your computer system, use the following steps to be certain the grammar and spelling checker is ready to serve you.

To use the grammar and spelling checker as you create your document

1. Go to Tools on the top icon bar.
2. Click on **OPTIONS** at the bottom of the Tools menu.
3. Click on the **SPELLING & GRAMMAR** tab at the top of the dialog box.
4. In the Spelling List, check all but the second and fourth boxes, which say

> - Hide spelling errors
> - Suggest from main dictionary only

5. In the Grammar List, there are more options than the four you see on the screen.
6. Again, check all but the second and fourth boxes, which say

> - Hide grammatical errors
> - Show readability statistics

7. Next, click on **SETTINGS** and then select the grammar options of concern to you.

As you type, a green line will appear under sentences that are considered to have grammatical errors, and a red line will appear under misspelled words. If you are using a PC, you can identify the grammar or spelling error by right-button clicking on the green or red line.

You can also run the grammar and spelling checker for the entire document.

1. *Click on* **TOOLS.**
2. *Click on* **SPELLING & GRAMMAR.**

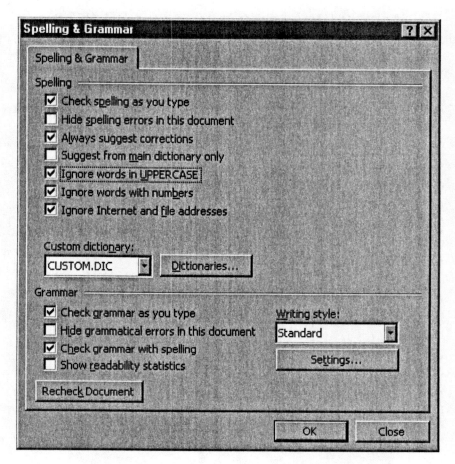

The primary settings for the grammar checker (Screen shot reprinted by permission from Microsoft Corporation.)

4. Now, apply the grammar checker to your version of the transportation engineering rough draft. Indicate your findings by using the editing symbols on your copy.

The Structural Principles of Sentence Logic

A sentence is a self-contained unit that should be designed so that its various internal elements share two logic structures: the inside pattern (of the sentence) and the outside pattern (of the group of sentences of a paragraph). The tools for composing the internal relationships for the parts of sentences are not very different from the tools we use for constructing the external relationships of the sentence groups. We use very simple logic processes—*coordination, subordination, parallelism, and causation*—to build and relate the elements of the component parts, both within a sentence and between sentences.

```
                SENTENCE LOGIC*
                Sentence Symbols
               Sentence Conventions
                   Word Rules
                  Number Rules
```

The key to appropriate sentence design is sentence logic. Any given statement you construct will call on a myriad of logical possibilities available to you in your language. Writing is not a simple system, such as basic math, in which a few functions and ten symbols get the job done. The latter uses a mere four processes:

$$+ \quad - \quad \times \quad \div$$

whereas the former uses a great many variants to perform the same tasks:

and	**but**
or	**for**
however	**therefore**
finally	**because**
nonetheless	**in addition**
moreover	**consequently**

It is important for engineering and engineering technology students to understand the similarity of the language and math processes. It helps to see that writing is not just the province of poets and that math is not just the comfort zone of engineers. In mathematics, the operational signs are called *verbs* and *conjunctions*. Any equation that involves any of the following symbols is using a logic process that is *identical* with words in the language system.

Verbs $< > \leq \geq \supset \subset \notin \cong \equiv = \neq$

Conjunctions $+ - \times = \cap \cup$

Look at writing as an analytical process, and try to write with a strong sense of structural organization derived from logic and from the logic systems of your engineering or techni-

---

* A boldface level is the focus of the chapter.

CHAPTER 2   THE BASE UNIT: LOGICAL SENTENCE

cal specialty. There are four basic principles you use in language organization. These basic tools are very important for building orderly, logical sentence structures.

Coordination

The first principle is *coordination*. Ideas are coordinated by a small number of specific words that place other words, phrases, or sentences in an appropriate balance in which one element is somehow matched with another or others. People coordinate language logic through their heavy dependence on the basic conjunctions:

and, or, but, for, nor

Larger units, such as sentences, are frequently coordinated (linked in a series circuit if you prefer) by the conjunctions:

> In terms of dendrochronology, the rings indicate that the tree was 300 years old, **and** one set of rings indicates that the tree survived a forest fire at the turn of the nineteenth century; **however,** for thirty years after the injury, as the rings indicate, the tree was seriously weakened.

Here you see that the first two sentences are coordinated by joining them with *and*. One of the larger connectors (conjunctive adverbs), which include the words *however* and *therefore,* joins the third sentence in the series.

Subordination

The second principle is *subordination,* which functions in much the same fashion as co-ordination. In grammatical terms, a subordinate clause is one that cannot stand by itself:

> Which is one possible cause of cancer.

Obviously, there is a *logical* as well as a grammatical reason why such a clause cannot stand alone. The preceding fragmented observation is dependent on, or is a condition of, another structure.

> The pollution included an emission of polychlorinated biphenyls (PCBs), which is one possible cause of cancer.

Here again the elements of the sentence are in balance. A logical process is called on by one of many terms used to initiate a subordinate idea:

after	before	unless	whether
although	if	until	which

because	since	when	while
	that	where	

The logic process demands a link to a preceding sentence. You will notice the logical utility of these words if you think in simple applications:

A before B

C unless D

E after F

If you recognize that the function signs of either system—mathematics or language—ask a reader to *execute* the function, you will see the secret to coordination and subordination:

$$A + B \quad \text{is the same as} \quad A \text{ and } B$$

Coordination and subordination are linking processes. Coordination tends to construct serial links among like units. Subordination usually constructs a before-and-after link, an if-then relationship, or some similar causal or dependent connection.

The external logic follows the same processes as the internal logic. If you are familiar with the principles for constructing an outline for a document or an investigation, you are familiar with the principles of coordination and subordination (see *Basic Composition Skills*, Chapter 9). Writers usually have more success with balanced and orderly organization if they outline a document before they compose it, because coordination and subordination are two basic logic functions that are used as organizers for language logic processing. You use exactly the same procedure in a paragraph.

The paragraph is a smaller unit of the larger process embodied in an outline, so it follows the same guidelines. It is important to realize that the *base unit* of this coordination and subordination process embodied in an outline is the sentence, which in turn is balanced by coordinating and subordinating the elements within the sentence.

Parallelism

A third structural principle is that of making coordinated units parallel. You must strive to balance the pieces of a sentence as though they were weights on a scale. This principle demands consistency in your logic, and it calls for balanced phrasing or *parallelism*, a type of duplication process.

Parallel structures in language are *like* structures. If you join two or more words, two or more phrases, or two or more sentences (independent clauses), you must balance the parts.

> Scientists have at last developed methods of "seeing" atoms, **and** one of these new methods is STM (scanning tunneling microscopy).

In other words, within the structure of even a simple sentence there are ways to compound the logic or serialize the thinking. You simply make parallel pairs or multiples out of words or phrases or clauses.

> The Dutch physicist Gabriel Fahrenheit proposed the Fahrenheit scale, the Swedish physicist Anders Celsius invented the Celsius scale, **and** the British physicist Lord Kelvin proposed the Kelvin scale.

You construct serial logic by connecting two or more items under discussion—*flora, fauna, concepts, chemicals, ideas, theories, species*—two or more of anything.

> The lab technician needs a pipette, a volumetric flask, **and** a centrifuge.

Of course, the series may be constructed of nouns, verbs, adjectives, similar phrases, and so on.

> Measuring, testing, **and** evaluating were the three tasks of the technicians.

The secret to handling a series correctly is to be sure that the first structure in the series is repeated in the rest of the series. This basic practice makes any pair or group clear to a reader. At times the *logic* is not parallel, as in the following sentence:

> Ø   She enjoys physics, math, chemistry, and **computers.**

The sentence should read

> She enjoys physics, math, chemistry, and **computer science.**

The logic of this list calls for four disciplines. A computer isn't a discipline, so to make the constructions logically parallel, the expression "computer science" must be used. Similarly, the *grammatical* structures must be parallel to maintain the correct logical structure of a sentence.

> Ø   The tasks of the FAA team involved **identifying** the structural problems, **representing** the problem in simulations, and **calculation** of the probable cause of the failure of the engine housing.

This sentence must be restated:

> The tasks of the FAA team involved **identifying** the structural problems, **representing** the problems in simulations, and **calculating** the probable cause of the failure of the engine housing.

The following sentence lacks balance:

> Ø The directions were simple: **attach** the clips to the red leads, **remove** the defective
> clips from the black leads, and the green lead is the ground.

This sentence should read

> The directions were simple: **attach** the clips to the red leads, **remove** the defective
> clips from the black leads, and **ground** the green lead.

Repetition is a device that is used extensively as a language-organizing activity. The repetition is not noticeable because the words are not repetitive; only the language structure is repeated. In contrast, if coordinated structures are *not* repetitive, the sentence logic becomes clumsy and very noticeably wrong.

Entire sentences are easily joined in parallel construction. Begin with a basic sentence:

> ENIAC was the first electronic digital computer.

You can add another simple sentence to the original. You duplicate the first structure to make both of them "parallel." Then you can put the sentences in a compound relationship with a conjunction.

> ENIAC was the first electronic digital computer, **and** it was completed in 1946.

> The computer consumed 150,000 watts of electricity, **but** solid-state technology
> would subsequently solve the energy problem.

Usually the parallel structure helps the reader see the series in equal units. A structure that is not parallel is often misleading in some way, or it may simply be clumsy.

> Ø To own or purchase a handgun you must be at least twenty-one years of age, a
> United States citizen, and a valid driver's license.

This sentence should be restated:

> To own or purchase a handgun you must be at least twenty-one years of age, you
> must be a United States citizen, and you must have a valid driver's license.

Here are a few practical rules that will help you identify situations that call for parallel logic.

- Look for the elements of a series if you used the words *and, but, or*. People tend to use these simple conjunctions for parallel structures within a sentence, whereas they often move on to the larger connectors—*however, therefore*—if they take the series beyond the simple sentence.

- Add words to create equal units in the structure of a series consisting of phrases.

  Ø   Drilling for pilings, the rebar structures, and pouring the foundation will be the most difficult tasks.

This sentence should read

> Drilling for pilings, **constructing the rebar structures,** and pouring the foundation will be the most difficult tasks.

- Change words to create equal units if a series of adjectives, nouns, or verbs is broken by a mismatch. Make the structure parallel.

  Ø   Dry copiers, laser printers, and **scanning** are excellent tools for technical writing.

This sentence is easily corrected:

> Dry copiers, laser printers, and **scanners** are excellent tools for technical writing.

- Comparisons should also consist of equal units. The signal will often be a word pair that compares. There are two words involved in the structure of such comparisons:

  either . . . **or**

  neither . . . **nor**

  both . . . **and**

  whether . . . **or**

  > *Both* the **database** *and* the programming **procedure** were problems that showed up in the program.

The points of comparison must then be structured alike.

  Ø   Both **the aqualung** and **diving with a snorkel** are popular for barrier reef research.

This sentence must be restated.

> Both **the aqualung** and **the snorkel** are popular for barrier reef research.

- Of course, you will be alert to a series because of the conjunction, but two additional comparison words call for parallel construction:

than

as

Most scientists find out that **earning** fame is easier *than* **living** up to public expectations.

Causation

When you construct a simple sentence, the grammatical term *simple* vividly defines the single-clause reality of the construction.

The star collapsed.

More elaborate sentences are referred to as *compound, complex,* and *compound-complex.* What is the logical point of other varieties of sentences? These structures serve the purpose of more complicated logic processing—and of causation in particular. Most reasoning is directed not at cause or at effect but at the connection of the two. This pursuit of causation is part of the scientific process, and it is no coincidence that the pursuit is part of sentence complexity. Sentences become complicated because sentences explore. They offer problems *and* solutions; they offer causes for effects and effects for causes.

The star collapsed **because** of the decline of its thermonuclear engine.

The rules for language usage are conventions or practices that have evolved to meet more than communication needs. The standards evolved to meet the needs for precision and accuracy for a logical process. It is for this reason that the language is a system akin to math systems. Every five or ten or twenty words or so you ask a listener or a reader to execute a function when you link logical units.

The star collapsed **because** of the decline of its thermonuclear engines, **which** is a process that begins when stars are several billion years old.

In a general sense, parallelism is the helpmate of coordination. Similarly, causation is often the helpmate of subordination. The four simple principles effectively work together to allow you to complete remarkably complex observations as well as the simplest of dialogs.

Sentence Types

There are four basic types of sentences. Most authors will mix and match the four as they are needed in the discussion of any given subject matter because the logic of each sentence is controlled by the four logic processes (see also pp. 102–105).

simple sentence (one independent clause)

He configured the position of the diode.

compound sentence (two independent clauses)

He configured the position of the diode, **and** he then reassessed the model.

complex sentence (one parent sentence and one subordinate clause)

He reconfigured the position of the diode **because** the earlier model was malfunctioning.

compound-complex sentence (two independent clauses and one or more subordinate clauses)

He reconfigured the position of the diode, **and** he then reassessed the model **because** the earlier version was malfunctioning.

Notice that the basic simple sentence is the root structure. The more complicated structures are the direct result of putting to work the four basic structural principles discussed earlier. The logical demands of the subject and the perceptions about the subject create the need for the various sentence types.

You should not have to make a conscious effort to blend these sentence varieties. Writing is not a recipe: writers do not use three simples and a complex and then add three more simples and a complex. The mix will be quite natural for most authors because they allow the subject to drive the expression of the discussion. Authors who are overly concerned with simplicity or complexity are tending to impose conditions on the subject of discussion.

The exception to this practice is the special focus you must always have on the reader. In writing technical communications, you must be consistently aware of what writing style the reader will need in order to understand the documents you are writing. You cannot simply assume that readers will have to adapt to your writing. You often have to adapt to them—in vocabulary, sentence length, depth of discussion, and so on.

Exercises Chapter 2, Section A

1. Coordination:

In the following passage, add conjunctions to join parts of sentences if the sentences seem logical as a single unit.

The lure of South America continued to attract scientists well into the nineteenth century. The German scientist Frederick Humboldt conducted a five-year, seagoing research mission that, significantly, began in the last years of the eighteenth century. The voyage ended in 1803, at the beginning of a new era. His mission was purely scientific. In his lifetime he was much venerated for his work. Charles Darwin was to be tempted by the South American continent as well. His five-year scientific voyage on the HMS Beagle began in 1831.

2. Subordination:

Use the following selection of words to help you subordinate several of the independent clauses in the following passage: after, before, unless, although, which, because, since, when, while, if.

Scientific reasoning began to assume entertainment value in the nineteenth century. The works of Jules Verne took science in one direction. The works of an American writer, Edgar Allen Poe, took science in another direction. Poe's stories involved the new science of criminology. It is no coincidence that the emergence of large industrial cities coincided with the need for policing ourselves. Indeed, the first director of the Paris Sûreté crime bureau, one Francois Vidocq, was on the inside track. He was himself a criminal, reformed. Edgar Allen Poe wrote the first detective stories. They were about an early French investigator who was imagined to be a student of the new science of criminology. These stories were set in France and appeared in the 1840s. This was decades before Holmes and Watson set to work with their magnifying glasses. Mystery stories reflect the scientific process reduced (or raised) to the level of an entertainment. Magnifying glass *in hand, we solve the mystery. The clues are the evidence. The process? It is inductive science, my dear Watson.*

---

3. Parallelism:

Look for sections of the following passage that could be organized in parallel structures and revise accordingly.

The microprocessor division is growing rapidly. It will develop at a pace well beyond all other production lines. Current production distribution involves guidance systems and tracking systems and the assembly of computers. The growth expectation in microhardware is creating a boom in microprocessor production, instrumentation, and researching microprocessors.

---

4. Causation:

In the following passage, add the appropriate phrasing to make appropriate causal relationships.

One of the crucial aspects of evaluating a touch screen's quality is the software application used to run the device. If the visual feedback to the user is not immediate, the user may get frustrated. The user will begin touching a single option repeatedly. This habit creates wear. Technical problems start. Another feature of a good touch screen application is that commands do not execute until the user releases his or her finger from the screen. This feature helps to ensure accuracy that will not corrupt user choices. Are there unsuccessful touch screens? Unfortunately, yes. Many touch screen applications are clumsily manufactured. The applications are not "user-friendly." Many lack a pattern or theme, which helps users finish their task quickly and easily.

---

5. In the margin of the following text identify the mix of sentence types that are used. The four sentence types are simple, compound, complex, and compound-complex.

Scientific treks are an old and venerable tradition. The intention of exploring new or little-known worlds has always tempted scientific investigators, and the desire to explore the unknown inspired some of the European voyages to the new world and the southern latitudes of the Pacific Ocean. In the late eighteenth century, Carlos III, King of Spain, sponsored three voyages, and his intention was to gather medicinal plants from the new world—and from South American in particular. Each voyage had a clearly defined mission, and the second voyage, departing from Europe in 1774, carried three botanists because this was a serious scientific voyage.

The idea of exploring was almost inevitably combined with the intention of returning with precious rarities, and the explorers often looked for fabled "treasures" also. For explorers with scientific interests, the plundering that sometimes resulted was discrete and intellectually motivated in some respects. The self-styled "archaeologists" of the nineteenth century provided Europe's museums with tons upon tons of "specimens," usually marble artifacts, because the explorers had financial support or personal wealth.

The Elgin marbles in the British Museum are from the Parthenon; they were bought in 1816, once they were already in London, which is a point of resentment to this day in Greece. The Pergamon Frieze was removed from Turkey, and it was shipped to Berlin in the 1870s. It took five hundred specially constructed shipping crates to haul the stones to Europe. Some of these artifacts were shipped out as crown "gifts" we are to believe; however, crown gifts are questionable if the royalty is a "colony" of another country. The seventy-five-foot obelisk in the Place de la Concorde in Paris was a "gift"; it was removed from the Temple of Luxor in Egypt. It weighs two hundred and twenty tons, which was quite a package for delivery in 1829. Logs of scientific investigations and illustrations to support the research were more practical tools because they took up far less space and involved little plundering and exploitation.

Making Sense of Objectivity

A larger issue—the desire to be precise and correct—defines many of the concerns you will examine in the following chapters. As technicians and engineers you might look at all the conventions in the *Writer's Handbook* and in other rule books as somebody else's grammar rules, but it is important to realize that your writing is driven by your professional objectives *and* by the rules. In the sciences and in the engineering and technical fields, the success of every mission depends on precision and accuracy. First, you must fulfill your commitments as an engineer or engineering technician in any project development. Second, you must provide the historical record of the proceedings or strategies or findings of your investigation or engineering application by documenting them. That documentation must also be precise and accurate if it is to render a scientific or technical process correctly.

You look at engineering drawings and never question the need for precision. You analyze the calculations for a proposal and realize the need for absolute accuracy. Writing is simply another element of the rigors of scientific procedure.

Objective Language

One aspect of scientific procedure that is important in formal writing is objectivity. A perspective that is objective assists in the design of logical sentences. Technical writing is quite formal and usually calls for the strictest objectivity. The word *objectivity* means "without bias of any kind." To be objective means to be open and honest and without personal persuasions. By contrast, to be *subjective* means to "take it personally." The subjective perception allows the personality of the writer to play a role in the writing, but the integrity of scientific investigation long ago dismissed this approach. Analytic reasoning depends on objectivity because the world around you reveals its own truths, which you then pursue as an investigator in your field of specialization.

You may not realize that many of the rules for correct writing reflect the objective method, such as avoiding slang and idioms to maintain language precision, and avoiding any language that slants the objective perception. If you disagree with a researcher, you do not call the fellow an "old crank" (at least not in writing). Instead you must demonstrate the failure of his findings so that you objectively prove he was in error. The demonstration will utilize the tools of coordination and subordination, causation and parallelism, and sentence variety.

Word choice is a key consideration in maintaining objectivity, as subtleties of meanings are projected by specific words. Consider, for example, a group of words meaning "overweight" or a group of words meaning "thin." Notice that they can be placed on a scale that runs from nice to mean if I approximate an order for them:

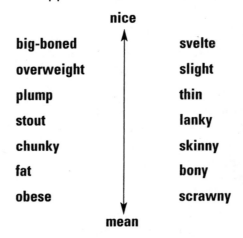

nice

big-boned	svelte
overweight	slight
plump	thin
stout	lanky
chunky	skinny
fat	bony
obese	scrawny

mean

The language is very precise; words can be graceful or rude. You can eliminate any suggestive tilt in your word choice when you want to maintain the objective position. Any term that is a tone-neutral expression is usually the "dictionary word" or what is called a *denotation*. *Overweight* is the denotation for weighing too much. *Obese* should also be denotative, but when you hear someone snarl the word accusingly you realize it is very slanted. *Obese* has a negative connotation. Any word that has emotional slant is said to have a *connotation*.

The words a writer selects create the tone of a document. Formal style employs a neutral tone in most business writing. Otherwise, the documents seem odd and unconventional.

Authorial Absence

Yet another aspect of objectivity is *authorial absence*. I occasionally mention myself and use the first person personal pronoun *I* in my writing. You have probably been taught to avoid writing with reference to yourself. This is standard procedure in formal writing, but it is important to recognize the utility of the practice. Not using the first person suggests an objective manner and improves the objectivity of a document. Certainly the *I* can be intrusive, and it can make readers a little skeptical of findings that are tinged with a researcher's ego.

Granted, no author would put pen to paper if there were not a personal drive motivating his or her efforts, but the absent-author style helps keep a document looking neatly objective. It can be argued, however, that the convention genuinely *does* create more objectivity in a document for a specific reason related to logic. Objectivity creates much more rigor in any comments an author constructs. If the *I* is removed, the logic stands alone. Consider the following sentences:

I think ghosts exist.

Ghosts exist.

If you were forced to select one of these sentences as true, or at least truer than the other, which would you select? Most observers would choose the first observation with the comment, "Well, the first idea may be true for whoever said it, but the second one simply is not true."

Speakers often preface their ideas with some variation or other of the *I think . . .* construction. They realize that they cannot be challenged because listeners will respect what other people believe—or at least they will be courteous about it. This tactic is at once defensive and *regressive.* The structure is regressive in that the logic falls back on itself:

I think ghosts exist.

The logic of the comment depends only on the speaker's *belief.* The objective style of writing demands that an author make an effort to *demonstrate* or explain a perception.

I think ghosts exist because some case studies have scientific validity.

There is now a little more logical material to discuss. The thinking is now what you might call *progressive.* The perception, once stated, is explored. You are forced to use logical analysis, usually causation, to defend your perceptions. This is the most significant value of the absent-author style.

Authors often use the *first person* form to indicate conjecture when they want to suggest a tentative position. Granted, the truth of any *I think* comment is relative, and, yes, the removal of the *I* can make comments much too absolute. Nonetheless, you need only insert words such as *perhaps* or *possibly* to indicate conjecture. The result becomes scientifically logical.

I think UFOs exist.	**(relative, tentative)**
UFOs exist.	**(absolute)**
Perhaps UFOs exist.	**(analytical, logical)**

Company documents typical of what you would write for a corporation are usually composed in the absent-author fashion. Most research reporting is equally objective, but you will see authors "enter" the documents from time to time. You can see this practice in trade journals in your field or in popular magazines such as *Scientific American,* most notably in the introductions. A scientific paper is very likely to open with a historical or somewhat anecdotal style in which the authors refer to themselves and their teams. If the text contains occasional similar self-references, you will observe that they are usually historical.

We arrived in Zambia with all of the solar eclipse equipment intact.

The discussion of the research will exclude the author's references to *I* or *we* to keep the findings thoroughly objective.

You can enter the document to set the stage:

In the following proposal, I will outline . . .

Educators discourage this phrasing because it is rather mechanical, even though there is heavy use of the technique for "papers" in industry. The method is perfectly acceptable if it helps an author get the job done—particularly for interim reports. If you set the stage by walking out in the footlights, limit the "*I* style" to your introduction. Also, be sure it is appropriate to your document and acceptable to your supervisor.

Exercises Chapter 2, Section B

Revise the following text so that it maintains an objective style and an objective viewpoint but retains the observations and position of the author.

There was a time when I could rely on a single Website search for all of my needs. Now there are more than 150 million pages of text and graphics on the Web, and finding relevant information can be a bear. The problem is relevancy! With the many search engines out there, the most desirable information is often buried deep in the "results list." As I see it, there is also a lack of quality information, and where there is quality, it is often available only for a price! So, how does one know where to go to get what he or she is looking for?

Excite is the best search engine for less experienced users. I have found that simple searches return good results, but advanced searches are ho-hum when you compare them with Yahoo! or HotBot. Excite also has trouble with symbols, such as "401 (k) plan." Excite does not return news results as well as other engines. The directory is broad—from animals to zoos—but I think it has an awkward feel when using it. The interface is beautiful and includes current stock quotes, general news, and chat calendars. However, the overall performance is just average. Some additional treats include free e-mail, chat rooms, personalization and customization.

AltaVista was originally brought to you by Digital Corporation. With all of that money in the bank, at least they could have done a better job with AltaVista. I have used this search engine for about three years and it does not compete with other engines such as HotBot. For me, the simplicity is inviting. The simple, straightforward approach to searching makes this site nice and easy to use. The results are fairly relevant and current. The interface is nicely organized and readable, probably because all of that weather, stocks, and headline garbage isn't there! AltaVista does give free e-mail accounts.

Viewpoint Problems

Shifts in Point of View

A number of minor problems affect the logical viewpoint of a document. The first of these problems is a point-of-view shift. You may not be generally aware of the text viewpoint when you read a document, but if there is a viewpoint shift you will notice the problem immediately.

> **She** studied the lower levels of the rock formation to identify any fossil remains. **Her** hopes that seashell remnants could date the structure were well rewarded when she located a number of shell fragments. **You** can then date the shells by carbon dating or by direct identification if the fragment is large enough.

The shift from a discussion of a female researcher to a generalized comment addressed to you, the reader, is awkward. The author should either continue with the researcher's activities

> She could then date the shells by carbon dating . . .

or else generalize the sentence without direct pronoun references:

> The shells could then be dated by carbon analysis . . . .

A writing project is usually composed in one of three viewpoints. You have the option of using any of the three viewpoints, but select one with care, and be consistent with it for your project. You can shift from one perspective to another, but the shift must be logical to the reader.

- ***first-person perspective***

  the *I* **viewpoint, including** *we* **comments**

- ***second-person perspective***

  the *you* **viewpoint**

- ***third-person perspective***

  the *he, she, it* **viewpoint, including** *they* **comments**

When you develop a project, use the perspective that serves the subject, the conventions of your company, and standard practices. For example, memos are usually composed from the first-person perspective, but manuals are not. If the manuals are directives, the document might call for the second-person or "you" perspective, particularly if an effort is being made to avoid passive sentence construction (see pages 117–119, 178–179).

> The badge color should be checked when leaving the chamber.   (passive)

This sentence can be composed to say

> You should check your badge color when you leave the chamber.   (active)

The third-person or "he or she" perspective is the least personal approach and is the strategy usually preferred in formal writing. Most of the writing output of a company shows no clear presence of an author, and this is particularly true of the wide assortment of pamphlets, booklets, brochures, manuals, and other hardcopy products generated as support tools for whatever services or products are generated by a business. Select the most appropriate viewpoint and be consistent in its application.

Number Shifts

There are several other viewpoint problems that are simple mechanical matters. The logic of the viewpoint will occasionally shift because an author changes the subject from singular to plural or from plural to singular. The viewpoint isn't actually changing; rather, there is a simple error in logic because one of something suddenly multiplied.

> A pharmacist's responsibilities have shifted dramatically in recent years, and **they** are now more concerned with patient risks and insurance liabilities since all the drugs are prepared by pharmaceutical companies.

This is an "agreement" problem. The author has to decide how many pharmacists are under discussion. Then, all the pronouns and verbs must agree with that number. The author can have one pharmacist (singular) or more than one (plural). The language and the logic must accurately reflect this basic fact. The distance between the noun and the pronoun (the number of words separating them) often determines the frequency of the error. The many words of our sentences can camouflage perfectly obvious logical connections, including such simple matters as a headcount.

Questioning the Reader

Finally, directly addressing the reader with a question or questions is seldom an acceptable practice, except in the case of direct correspondence.

> Could **you** have improved the accuracy of the data?

Questions have very limited use in formal reports. They can build interest in introductions because they are provocative, and they can be used as paragraph transitions or section transitions. Transitions are words, phrases, or sentences that connect the parts of a document. Transitions usually are positioned in the first sentence of a paragraph. They are most important at any point where the logical progression is not self-evident. Questions can be helpful in this regard.

Using questions in conclusions may be risky, since questions call for answers, and you may not want to open avenues of investigation you did not pursue. If you do construct questions, address the issue and *not* the reader. Notice that the following question is directed toward the reader and *not* the subject, even though the word *you* is removed.

> Could the accuracy of the data have been improved?

The question can be placed in a context that will not point at the reader.

> The outcomes of the report provoked many questions. First, could the accuracy of the data have been improved?

Notice how easily this question constructed a transition from the discussion of outcomes to the issue of accuracy.

Exercises Chapter 2, Section C

Revise the following text so that the point of view is consistent and the pronoun usage is clear.

Early explorers neither understood the risks nor perceived the ethical issues involved in gathering specimens of interest to the European scientific community. Granted, I realize that the plants that grace today's gardens have been gathered from throughout the world. Indeed, amateurs and professionals alike created extraordinary collections of seeds, living specimens, and carcasses of everything from bats to bugs to bananas. You can imagine the importance of the collections. The scientist who gathered enough of anything felt that he could contribute to our knowledge of the world in some way.

Fortunately the temptation to "bring 'em back alive"—or dead—was dimmed by the realities of storage and maintenance for specimens. Although the British Museum had been created by an Act of Parliament in 1753, the natural history department was a notorious mess until the mid-nineteenth century. They were very poorly organized. In addition, in the 1840s, the American Phineas T. Barnum began his long road to making a three-ring circus out of specimens, alive and dead, which may have discouraged any scientist from appearing to be involved in any such commerce—with exotic animals in particular. You need a safe and ethical approach to the retrieval of specimens.

Safety, in particular, has become a major concern for scientists, I'm certain. The sparrow, the Japanese beetle, the killer bee—these represent one variety of risk. However, people represent a risk in themselves. As carriers of disease, people took smallpox into the new world, and the vengeance of a new world scourge, syphilis, was taken back to Europe. Then there was the introduction of tobacco! Our first landing on the moon posed a serious question concerning contamination. How safe do you think it would be to return with lunar specimens? The decontamination of the returning moon landing crew and their lunar specimens was as thorough as possible, but only within the constraints of our scientific know-how. The risk is always there. The astronaut can bring back unwanted problems if he is not properly decontaminated.

Style and Word Choice

Style is an aspect of sentence logic that is often overlooked. Almost everything you write should be formal in style except perhaps for casual in-house memos. Since you will represent a company and a company image (be it your own business or a large company), you need to abide by the standards of the industry. Formal writing—whether in law, science, engineering, or the technologies—must be precise, and there are practices you can use in your writing to create precision.

Formal writing is easily identified. It is neutral in tone, as you have observed. It is highly polite. It is discrete rather than blunt. It is *detailed*. The details lend precision to the logic of the writing, but the details must, in turn, be handled precisely. Compare the following two examples:

> I'm writing to ask for a bid to have a lab with twelve sinks plumbed in to spec with lab grade fixtures. We would like your bid by June 12 because . . . .

> C and C Architectural Subcontractors is presently seeking job estimates for the plumbing subcontract of a new laboratory. The laboratory will need lab grade fittings and drain lines rated to meet the corrosion levels identified in the specifications attached. We hope to have the contract proposals by June 12 in order to meet the general contractor's schedule.

As you see, the key to politeness and a discrete approach is primarily the method of word choice. Notice the thoroughness of the second sample. Thoroughness is a logical necessity. The issue is *not,* as some writers might think, "big words" to replace little words. You do not need pretentious language. You simply need to be polite and thorough.

Informal style is the style of conversation; formal style is the style of business writing and technical writing. If you eliminate conversational expressions, such as *well,* and most of your contractions (*don't, can't*), you are well on your way to the style that is appropriate.

Specific Nouns and Verbs

Formal writing is very precise. It is important to select the most precise nouns and verbs to convey the correct information to the reader. A precise vocabulary will inform a reader more accurately than generalized nouns or verbs. Adjectives and adverbs can then be added to further describe the terms.

horse	**write**
palomino	**write a memo**
palomino gelding	**write a memo quickly**

Technical writing does not depend on the adjectives and adverbs that usually make writing colorful. Technical documents tend to focus on nouns and verbs and logic links. Precision in nouns depends on scientific terminology, brand names, product types, and code numbers. The following figure illustrates six ways to say the same thing, depending on the precision required.

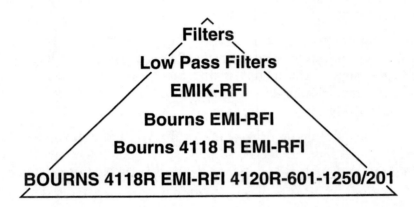

In technical documentation, adverbs are used to precisely define verbs. Adverbs are the usual descriptors that tell us more about a verb or action, usually by answering the questions *When? Where? How?* I might say that a task is subsequent to another or that it is performed in a specific location or that it must be done *gently* or *never* (if it is dangerous).

> This task comes **later.**
>
> I fly **often.**
>
> We perform this task **here.**
>
> It fell **rapidly.**
>
> The tests went **badly.**
>
> Proceed **slowly.**

As you can see, an adverb can detail an action within a sentence. A set of instructions performs the same mission on a larger scale. Technical documentation is frequently an enumeration of tasks that explains *when* to do them (and in what order). Assembly drawings are sprinkled with legend numbers that can identify *where* to do the job. The motive for the entire process explains *why* there is a precisely defined strategy. The strategy itself explains *how* to achieve the ends of the process. It is important that the language of a technical document describes these tasks with equal precision.

Readers of technical writing are accustomed to precision. Precision is the production criterion for the best product, whether it is the assembly, the assembly drawing, or the assembly instructions. The neutral style is one aspect of our goals, and precise word choice is another.

Jargon

The specialized vocabulary that is used by any particular occupation, profession, or discipline is called *jargon*. It is tech talk. Engineers, scientists, and technicians could hardly get along without the curious mix of slang—*boot up the system, he smoked the circuit*—and precise language—*trihedral prism, prototrophs, chromatograph, sequential video switcher*—that characterizes science and engineering fields.

As a vocabulary for communication, however, jargon presents a problem. In spite of the precision that jargon brings to a technical field, it is a closed circuit that eliminates the uninitiated. In developing any technical document, determine the likelihood that your readers will understand the jargon you plan to use, and use it accordingly so that you communicate effectively.

Computer-related technology is one source of the vocabulary problem. Discussions of programming languages, computer technology, software applications, hardware and network architectures, and Internet computing can be a maze of jargon. Here are a few of the typical terms:

C++	Computer Technology	Internet Communications
.h file	command-line interface	uninstall utility
loop invariant	multiprocessor OS	file transfer protocol

To make matters worse, many terms are corruptions of or borrowings from English, which are often confusing to the reader who is unfamiliar with the material. Here are just a few terms that have been recycled:

C++	Computer Technology	Internet Communications	
compile	icon	compress	save images
modular	kernel	token	gopher
	prompt	tree list	thread
	menu	hide windows	hit
	server	join tables	

Because much of this technical vocabulary is unavoidable, it might be helpful in programming and computing documents to set the terminology in italics or boldface, either at the first occurrence or in every instance if the product is to be read by a novice. (See also pp. 82–85.)

> A *word value* of 1234h in this location is a *flag* that indicates a *warmstart,* which causes the memory test portion of the POST (Power On Self Test) to be skipped.

Slang

> Hey, like you know, that band was radical. So cool. But man, the dude beside me needed to chill out.

I realize there is no chance that you could mistake this sort of language for the concrete and precise language of formal writing.

Clutter

Most writing manuals discuss other forms of language that are inappropriate in formal writing. Because technical writing tends to be scientific, authors of technical documents probably have less of a tendency to use regionalisms (such as country talk) and clichés and obvious idioms. You will not find a surveyor saying, "Oh here back about three mile" or a megafarm owner writing annual reports to stockholders with expressions such as "sitting tall in my saddle" (meaning *proud*) or "looked like he swallered his chaw" (meaning *shocked*) or "we howdied awhile" (meaning *talked*).

Idioms, however, are another matter. They are very common. Some are obvious, and some are not. An idiom is any word or group of words that say one thing but mean something else. The ultrasound technician about to use the ultrasound scanner probe says to her assistant,

> Keep an **eye out.** See if you can **catch** the image.

She means, of course,

> Watch carefully, and be prepared to look at the image on the screen.

Idioms can be very colorful, but they can take on the characteristics of slang:

> The patient was thrilled to pieces because the test was negative.

> The patient came unglued because the test was positive.

Idioms are not always this obvious, which is the problem. The occasional idioms that find their way into technical writing are usually recognizable because they frequently end *up* with a preposition: *call up, wind up, blew up, take up, stir up, wrap up, fueled up, fired up, fouled up, tune up, knock down, turn down, cut down, run down, wind down, ran across, put up with, fade out.* Watch *out* for these and others.

> He **ran up** a bill.

> The supervisor **handed in** his resignation.

Popular Appeal

There is nothing wrong with writing in a popular or conversational style that is relaxed and appealing. There is a large industry of technical trade magazines that seek a non-threatening style as a way of making technical articles as user-friendly as possible. There are, however, two reasons to handle popular writing techniques with caution. A magazine writer's style is usually much less formal than a corporate style, and the magazine author has a unique audience that you would not usually address in writing. In addition, in recent years the method has become much imitated, and the phrasing tricks have reached the point that expressions have become clichés:

> It really isn't complicated, **so what's the catch?**

> There are a lot of integrated circuits **out there.**

> This model is not your **ordinary garden variety.**

Language Politics and Gender Neutrality

Sexist language is a public issue that was addressed in the 1960s. In response to the sexism that was identified in writing, business and industry adopted new practices for developing documentation that was gender neutral. Standard practices in business correspondence needed updating, and company manuals needed stylistic revisions to reflect workplace realities.

Well into the 1970s it was still commonplace to see a letter open with *Dear Sir:* or *Gentlemen:*. These were usually generic references to no one in particular. The work force is now 55% female, and so the unrealistic use of male generic references—as well as the gender politics of these expressions—needed attention. The use of *Ms.* to address women was a particular breath of fresh air because the choice between the older, awkward *Miss* and *Mrs.* was seldom clear.

For most writers, the politics has not been an issue; the problem has been a language difficulty they need to analyze. In a wide assortment of technical documentation—and in assembly manuals in particular—the long-standing tradition was to use *he* as a generic reference to the assembler. Today's authors are aware of the gender bias of the generic *he* and avoid it in writing. The grammatical glitch concerns the well-meaning results of gender neutrality. It is easy to remove *manpower, workman,* and the verb *to man* (to operate), but getting rid of *he* can be a headache for writers. One option is the expression *he/she.* In a pinch this is satisfactory, but it is not exactly good English. The slash was never intended for hyphenated tasks, and the construction reads oddly and, if spoken, does not make much sense.

The option is to omit the singular pronoun group (*he, she*) and use plural pronouns and nouns instead. In other words, if a manual referred to a *he,* you simply replace the *he* with *the technician* or use the plural form *technicians* and the plural pronouns *they* and *them* (see p. 197). *You must, however, be consistent.*

If you shift from one assembler to more than one, do not do it in the same sentence. If the shift occurs within a sentence, the logic is clumsy; you will then have two or more people where you had only one a moment ago, and the result is a grammar error in agreement. Spoken versions of this error go unnoticed:

Ø   If **somebody** calls, get **their** number.

In writing, the error will usually be more obvious, and this glitch is one of the most common errors in writing of recent years:

Ø   The skilled network **designer** should know **their** latest application architecture for success in properly installed systems.

The preceding error is particularly easy to correct, since the sentence did not need the personal reference, *their*.

The skilled network designer should know **the** latest application architecture for success in properly installed systems.

Here is another example:

Ø   The **alcoholic** has a much greater chance of killing someone while driving than any other person with an illness of any other kind. **They** are quite reckless when **they** drive.

The error in this sample is easily corrected if you make the reference to the alcoholic plural.

Alcoholics have a much greater chance of killing someone while driving than any other person with an illness of any other kind. **They** are quite reckless when **they** drive.

Make sure you shift the logic of your head count if you want to use *they* and *them*. Shift to a plural noun first to reestablish the logic for your readers.

Let's run through an example. If I am making a general reference to an assembler, I now avoid the historical practice of using the pronoun *he*.

Ø   **He** then dismantles the drive unit.

The easy solution would be to replace the word *he* with the word *one*, but this practice is out of date and seldom used.

One then dismantles the drive unit.

Notice that the result sounds stuffy and old-fashioned. In order to reconstruct the sentence, I use a noun, the *assembler,* or move to plural pronouns, *they* or *them,* which show no gender.

> Noun form: The **assembler** dismantles the drive unit.
>
> Pronoun form: Then **they** reposition the wing assembly.

You can see that I jumped from one assembler to many assemblers. This is the problem authors must handle in gender-neutral writing. I *must* use a plural noun to initiate the use of the plural pronoun.

> The **assemblers** lower the wing assembly to the fuselage.
>
> **They** must be extremely cautious.

Note that the nouns can shift back and forth between singular and plural, and the commentary will usually appear to be perfectly logical, but the same is not true with the noun-to-pronoun transition:

> The **mechanical engineer** should register with the Sacramento Engineering Department by May 1.
>
> The **engineers** must not be late.

Nouns shift back and forth as you can see in the sample, but the reader's eye will not accept a similar shift in pronouns. Readers see the nouns as separate entities—*engineer* and *engineers*—but they link the pronoun to a noun and expect them to match.

You can, for example, construct this pair of sentences:

> The lab **tech** was exposed to a low-level radiation leak. The **technicians** were warned to carefully monitor their badges.

Notice the shift from singular to plural. A similar pronoun shift will not work properly:

> The lab **tech** was exposed to a low-level radiation leak. **They** were warned to carefully monitor their badges.

Of course, the use of *you* or an imperative sentence also avoids the gender problem:

> You can then align the three components.
>
> Measure the density first.

Gender Shifts

There is a somewhat popular effort to mix references to *he* and *she* when the pronouns refer to generic members of a group, of employees, for example. One paragraph might say,

> She should fill out application W-9.

Another paragraph might state,

> He will also need a J-4 form.

This practice is workable as long as readers will not be misled into assuming, for example, that certain procedures are only for women. The method is, however, a little clumsy as a technique for accommodating gender neutrality.

Exercises Chapter 2, Section D

Examine the following five passages. Make any necessary changes to produce more effective phrasing.

1. If the last two (signature) bytes of the master partition boot sector are not equal to 55AAh, software interrupt 18h (Int 18h) is invoked on most systems. On an IBM PS/2 system, a special character graphics message is displayed that depicts inserting a floppy disk in drive a and pressing the F1 key. For non-PS/2 systems made by IBM, an Int 18h executes the ROM BIOS-based cassette, BASIC Interpreter.

2. For starters, let's go over three drop-down menus for style, font, and font size. The style drop-down menu would let you run through the particular styles for headings and subheadings. The font drop-down menu would be used for picking out the font, and the font size menu would be used to wrap up the basic formation selection.

3. Excel is very popular and dynamite powerful because of its flexibility and useful functions. Especially, "Charts" in Excel has very sophisticated features, and there are many types of cool charts you can choose, depending on your purposes. You can draw almost any kind of graphic chart you can come up with. Drawing charts in Excel is a very friendly process, even for a beginner. Now, let's learn how to draw it!

4.   • *Type your message in the appropriate field.*

   • *Click on the Send Later button.*

   • *Sign on to AOL.*

   • *Click on the Mail menu.*

   • *Select Read Outgoing Flash Mail from the drop-down list.*

   • *Click the Send All button as soon as you see your message title displayed, and the program will send your mail in a few seconds.*

---

5.   *Adding scientific formulas to texts has become a relatively easy task for an engineering student. A complicated scientific notation takes only a few minutes to add to a text if he knows the correct way to do it. Using the MS Equation Editor is quite beneficial to him because it adds a professional look to his science and math projects.*

The Organizers

The punctuation that writers use in the process of writing consists of about a dozen basic marks that allow readers to see the starts and stops of text logic. These logic symbols are either "outer" or "inner" dividers of the logical sections constructed into a sentence or group of sentences. The outer marks indicate sentence stops. The sentence is coded to begin when the first word is capitalized. The unit of logic ends with one of the symbols that you know quite well:

.	**period**
?	**question mark**
!	**exclamation point**

Of course, in engineering and scientific fields most sentences will conclude with a simple period.

```
              Sentence Logic
          SENTENCE SYMBOLS
          Sentence Conventions
             Word Rules
            Number Rules
```

There is a second group of punctuation symbols that you need to handle with skill. These are the inner marks that subdivide features of the sentence logic. They include almost twice as many marks:

,	**comma**
;	**semicolon**
:	**colon**
—	**dash**
()	**parentheses**

The comma is the most versatile tool of all the punctuation marks and probably appears more often than all the other inner marks combined.

A few special devices are the underline, *italics*, and **boldface**. They "punctuate" any word or series of words, and they are particularly handy in technical documentation.

Several additional punctuation marks will be considered under a discussion of quotations. The marks that designate quotations and deletions (elisions) are the following three symbols:

	deletions
" "	**quotation marks**
' '	**single quotation marks**
. . .	**ellipsis**

Outer Punctuation

Let's look at the use of the three types of outer punctuation.

The Period

- End declarative or imperative sentences with a period.
- Do not use periods with lists composed of short or long *phrases*.
- Use periods with sentence outlines.
- Do *not* use periods with topic outlines.
- Do *not* use periods with capitalized acronyms.

ASCII	CAD	MUD

- Do *not* use periods in the popular abbreviations of technical devices or processes.

SLR	DNA	URL
MRI	CPR	TCP/IP

- Do *not* use periods in the popularly accepted abbreviations of companies or agencies.

IBM	NASA	GM

- Do *not* use a period after abbreviations for units of measure (for exceptions, see pp. 235–236).

$$40 \text{ cm} \times 20 \text{ cm}$$

The Question Mark

- End questions with a question mark.
- Do *not* use the question mark parenthetically to suggest that there is some question about your own ideas or the ideas of others.

  Ø The research (?) was just luck, since Fleming just happened to observe the behavior of penicillin in an unrelated experiment in 1928.

- Do *not* use a question mark in the middle of a sentence unless you can construct the sentence as a quotation.

  Ø There was no effort to ask, Should X-rays and radioactive radiations be regulated? until 1927, when the American biologist Hermann Muller observed that X-rays had caused insect mutations.

- Do *not* use more than one question mark in an effort to build emphasis.

  Ø   How could a microprocessor the size of a postage stamp perform 300,000 logical functions in a second????

The Exclamation Point

- End energetic comments with an exclamation point, but realize that its use is rare in scientific or engineering work. One of the many devices that writers use to control and sustain the objective style is the monotone of a neutral voice in which the zeal of any exclamation is avoided.

- Do *not* use more than one exclamation point to turn the volume even higher.

  Ø   The sun's surface has a temperature of 6000 K!!!

- Do *not* use an exclamation point parenthetically.

  Ø   The Ebola virus research (!) is very dangerous since the mortality rate can reach 90%.

Exercises Chapter 3, Section A

NOTE: At the end of the subsections of this and the following chapters, you will find brief editing exercises relating to punctuation, grammar, and similar matters. These exercises are not intended to be drills. Each exercise is designed as a comprehension check for each chapter subsection. Since many of the rules you will be reading about are critical, but often subtle, the exercises will help you focus on relevant details that are important in writing, particularly in technical communications. Try your hand at the exercises and consult the text for the rules.

If you repeatedly commit errors of any one kind, ask your instructor for a text that contains appropriate drills. The use of drills is a way to learn and reinforce skills through repetition.

Correct the errors in the following sentences.

---

1. *He worked for G.E. and then went with I.B.M.*

---

2. *An M.R.I. scan looks like an X-ray, but the technology is based on magnetic resonance imaging.*

---

3. The U.F.O. (?) turned out to be a unique form of electrical discharge called a "lightning ball."

_____

4. The lightning ball is a ball of white light a foot in diameter that can move parallel to the ground!!!!

_____

5. In the 1950s everyone was wondering, Why was the government not investigating the UFO sightings? and there was a national fear of the phenomena.

_____

The Workhorse: Comma Basics

Commas divide a variety of shorter features of a sentence. They are the workhorses of sentences. *Semicolons* and *colons* usually extend the logic of a sentence into a second sentence, which is conveniently joined to the first with one or the other mark. The *dash* and *parentheses* are used to identify secondary material that would otherwise create a clumsy blip in a sentence.

Math systems are remarkable tidy compared with language systems. Math systems use a fairly small group of signs to represent quantities and even fewer signs to represent functions or processes. By comparison, language is either sloppy or wonderfully flexible, depending on a writer's point of view. With so many options available to you—particularly in the way you construct or design the word order—you must subdivide the elements so that readers can work their way through your thinking.

Math systems are usually predictable, partly because they are symmetrically rendered. Writers try to make the language system orderly with such tools as the comma. As noted previously, the primary use of the comma is to subdivide. As words shape the precision of your meaning, the commas help separate those words into logical clusters. For this reason, the comma always follows or precedes words, phrases, and clauses.

There is a distinction between commas that set off elements of sentence logic and commas that set off elements that are secondary and auxiliary. The first group includes commas that surround words, phrases, or clauses without which the sentence fails to operate properly. The second group of commas helps to shape a host of other material an author is willing to include but, alas, does not exactly need in order to make the sentence function.

Key Commas
Compound Sentences

If you join two sentences with *and, or, but, for, yet, so,* or *nor*, place a comma before the connecting word (a conjunction).

> I watched the storm, **but** I knew I was in no danger.

Dependent Clauses

Use a comma after an introductory dependent clause.

Although there are 25,000 miles of blood vessels in the human body, the blockages that cause heart failure generally occur within inches of the heart.

Use commas to isolate a dependent clause in the middle of a sentence, unless it begins with the word *that*.

Canada and Brazil have approximately the same areas of forest, **if we compare boreal forest with rain forest,** but do not protect the same amount of forest lands.

This concept **that gas will occupy a smaller volume** was discovered by the geologists.

The comma is often omitted when a dependent clause is placed at the end of a sentence, except when the clause expresses contrast and begins with a conjunction like *although* or even *though*.

Conventional television has poor image quality in contrast to high-definition television **because HDTV has four times as many pixels to create an image.**

Teflon is a flourocarbon, **although it has unique characteristics.**

Transitional Adverbs

Use a comma or commas to set off a very important group of logic links that are used to move the reader along the logical path of a sentence. Some examples follow:

therefore, however, consequently, finally, similarly, subsequently, indeed, after all, perhaps, certainly, instead, nevertheless, nonetheless, thus, accordingly

The first five in particular are often used as part of paragraph transitions, but all of them can be used to provide the reader with a logical decision or logical transition.

The mass production of television sets began in 1947, because, **finally**, production costs were declining.

Costeau was convinced, **nonetheless**, about the design for the aqualung, which he first tested in 1943.

Series

If you construct a list within a sentence, use commas to separate the elements.

The most dangerous isotopes are U238, U236, and U234.

The last comma in a series is optional in the popular press, but the style needs to be consistent throughout any one piece of writing.

If the series involves inventory (materials of any kind), the list is quite simple to construct. If the list involves tasks, you will be joining phrases or clauses, which calls for a little more attention to detail. Observe that the commas distinguish each element in the series to identify the separate units. You can look at the function of the comma in two ways: the commas *join* parts but also *separate* and define the parameters of those parts.

> The emergency hatch is secure only if the support tech is certain that he or she mounts the frame bushings, attaches the Teflon cushions to the bushings, and cinches the housing nuts with the hatch torque wrench.

> Each member of the crew had specific responsibilities: the navigator was in charge of the flight path, the copilot handled the flight plan, and the pilot managed the actual flight of the craft.

Remember that it is common practice to avoid lists in the running text (in the body paragraphs) if you are developing a document in an engineering or scientific field. Technical writers prefer to align them in indented columns even when the lists are quite short.

Adjectives

When you use two or more adjectives to describe a noun, you use commas to separate the adjectives.

> The **foul, damp** atmosphere in the lab was a problem for the tests.

> The **large, bright** nova was more than apparent in the photograph.

> The **small, cold, metal** forceps bothered the patient.

Use hyphens to join word clusters that are read as a single adjective.

> Aspirin is a simple **over-the-counter** drug.

Support Commas

Because writers can shuffle the parts of a sentence, they can compose a sentence in a number of ways—as long as they use commas to mark the divisions. "Key commas" make important organization distinctions for a reader so that the reader sees the logical word groupings. "Support commas" perform a host of what might be called second-rank tasks that clarify a number of other constructions that are often used. In these secondary constructions, the omission of the comma will leave a noticeable blip or choppiness in the text, which is why commas serve cross purposes: to separate parts and to signal the connection at the same time. The commas are gates that batch the elements.

Introductory Words, Phrases, and Clauses

A variety of constructions can introduce a sentence to explain its details. Place commas at the end of introductory constructions so that the discrete parts will be clear to the reader. The reader will see the connection but will not be confused by the construction. The comma joins the component parts but also marks the distinct elements of the parts of the sentence.

Simple introductory words or phrases

Second, the popularity of the typewriter was gaining ground, and forty models were on the market by 1900.

In addition, air-conditioning systems greatly increased the energy demands in the rapidly growing Southwest.

Other constructions are more elaborate. Use a comma after prepositional phrases.

Introductory prepositional phrases

Under heavy pressure from competitors, nineteenth-century engineers struggled to mass produce glass products of all kinds.

In the period that coincided with the mass production of pharmaceuticals, Ashley developed an automatic bottle maker that could produce 2500 bottles per hour.

Introductory participial (verbal) phrases

Use a comma after long introductory phrases that animate a noun.

Risking life and limb, the volcanologist lowered himself into the crevice.

Torn by doubt, she did not want anyone to see her findings from the field trials.

Having overlooked the Canadian snow-load problem, the engineers greatly underrated the bearing load for the building.

Notice that many of these phrases are signaled by verbals ending in *ing* (see pp. 105–107).

Descriptive Phrases and Clauses

Adjectives are often constructed as a group of words, and the clusters may well contain subjects and verbs (adjective clauses). Adjectives usually come before a noun, but adjective phrases and adjective clauses can also be placed after the noun as you see here.

The pygmy, **though short in stature,** shows few signs of dwarfism.

The other tribes, **which were much smaller in number,** did not usually challenge the Mauri.

The stars, **at least the red giants,** are huge objects.

The stars, **including those that are in the process of their own destruction,** are billions of years old.

The hydrogen-filled dirigible, **with hardly any engine power at all,** was a slow and risky means of air travel.

Appositives

If you describe a person, place, or thing before or after you cite the actual name in question, you are probably going to need commas to offset the name or the description. These repetitions (called appositives) are rather like synonyms or analogs. They construct an equivalency but one that is choppy if you do not use commas.

The blue, **a cobalt color,** was intense.

Edwin Land, **an American inventor,** first marketed the Land Polaroid camera in 1948.

Patricia, **my colleague,** is an anthropologist.

The head of the German rocket program, **Werner von Braun,** fired the first guided missile in 1942.

In fact, many of these constructions are simple definitions and are ideal in technical writing.

A glitch, **an electronic transient,** will occur instantly.

Mud, **a premixed commercial grade of plaster,** is usually very light and manageable.

Antithetical Phrases

If you construct a negative clarification, set it off with commas.

Crick and Watson, **not Avery,** determined that DNA encodes genetic information.

Misunderstandings

Commas are commonly used to create any useful divisions that will help avoid confusion for readers.

The two researchers, Chargaff and Franklin, led the way in the first phase of DNA research.

Dialog Tags

Use a comma to conclude the introduction to a quote.

> Einstein commented, "Great spirits have always encountered violent opposition from mediocre minds."

Secondary Logic

A number of elements that are not necessary to the basic logic of a sentence are usually separated with commas.

Parenthetical concerns

Writers frequently include supplementary material in their sentences. If the material can be deleted, it is said to be parenthetical.

> John Snow, **who was the first to be suspicious,** realized that water was causing the spread of cholera in London.

> This incident, **for all practical purposes,** demonstrated that water carried diseases.

That, which

The word *that* is not preceded by a comma (because it is used to construct restrictive clauses).

> The findings **that** demonstrated the connection between water and disease could not be explained for several decades until Pasteur, Koch, and others began to understand bacteriological contamination.

The word *which* often takes a comma (because it is used to construct nonrestrictive clauses).

> One bacterial source was quickly identified, **which** helped Snow isolate a water pump as the possible cause of the disease.

"Which constructions" are nonrestrictive and are secondary to the sentence logic, and "that constructions" are restrictive or important to the sentence logic. In other words, the word *nonrestrictive* means *unnecessary,* and the word *restrictive* means *necessary.*

In the following sentence, for example, the use of the word *that* is part of a necessary construction.

> The car that I drove was too small.

Without the restrictive clause the sentence logic is altered.

> The car ~~that I drove~~ was too small.

However, in the following sample the word *which* does not begin a construction that contributes necessary logic to the sentence:

> The car, which was red, was too small.

Without the nonrestrictive clause the sentence continues to function.

> The car was too small.

Notice that the restrictive clause is necessary and is *not* set off with commas, whereas the nonrestrictive clause is not necessary and *is* set off with commas.

Titles

Use commas to separate a person's title from a person's name when either is used as an appositive.

> Alexander Houston, **water board director,** built a chlorinated water treatment facility in Lincoln, England, late in the nineteenth century.

> **The chief naval officer,** John Smith, was concerned about water contamination on ships.

Addresses

If an address is placed in a sentence, separate the elements with commas. The street address seldom appears in this context. If you identify only the city and state (or country), place commas before and after the state (or country).

> The first metropolitan installation of a chlorinated water treatment facility was constructed in Middlekerke, Belgium, in 1902.

Dates

In presenting a date, you may omit all commas if you work for a company or agency (military sectors, for example) that uses the following style:

> The first self-sustaining chain reaction began at 3:45 P.M. on 2 December 1942 at the University of Chicago.

Use commas to preserve the more conventional format using month, day, and year.

> On February 18, 1930, Clyde Tombaugh discovered the planet Pluto in photographs that were taken at Lowell's Observatory in Flagstaff, Arizona.

However, do not separate a month from the year.

> The data were collected in June 1999.

Exercises Chapter 3, Section B

Add the necessary punctuation to the following sentences.

1. The cardiac pacemaker weighs 1 to 5 ounces and a lithium battery provides enough power for the device to run from 2 to 15 years.

2. Ivan Pavlov used a bell to test his theory of conditioned response and his experiments became more complex once he understood the implications of his discovery.

3. Because too many of the NASA space vehicles were not reusable the agency launched the space shuttle program in 1972 with the idea that a craft in flight could safely return for a landing.

4. Although one shuttle was lost during the takeoff disaster of 1986 the other four proved subsequently that mission vehicles could return safely.

5. Lasers can read grocery barcodes clean clogged arteries and guide weapons to their destinations.

6. Alternative energy resources involve many natural sources of renewable power: wind tides geothermal heat and sunlight.

7. Vacuum tubes were large hot and unstable.

_____

8. A small blurry image showed up on the screen in blue red and green colors during the first color television broadcast in 1940.

_____

9. Under the research direction of Peter Carl Goldmark early color television technology was developed for CBS.

_____

10. However it would be several decades before the technology could be perfected.

_____

11. Deeply disturbed by the scourge of polio Jonas Salk and Albert Bruce Sabin succeeded in developing vaccines to control the virus in the 1950s.

_____

12. During the final phases of World War II antibiotics became part of the war effort and saved the lives of many Allied soldiers.

_____

13. Penicillin though more effective than sulfa drugs was not an immediate commercial success because additional research and experimentation were needed.

_____

14. Tuberculosis which resists penicillin was treated with the antibiotic streptomycin which was developed by Waksman in 1947.

_____

15. Morse an American inventor patented his telegraph in 1837.

_____

16. The German physicist Walther Müller collaborated with Hans Geiger on the development of the radiation counter which came to be known as the Geiger counter.

_____

17. The robot a mechanical humanoid was first conceived by Karel Câpek in 1921.

_____

18. WKCR the first FM station went on the air at Columbia University in New York City in 1941.

_____

19. One of the first computer bugs was a moth that shorted a relay in an early computer in 1945 and it was duly recorded in the logbook of the Bureau of Ordinance Computation Project at Harvard University "First actual case of bug being found" (The moth was also taped into the logbook.)

_____

20. Hoover marketed the first electric vacuum sweeper in 1908 which was just two years after GE released the first electric range.

_____

21. In October 1960 Jane Goodall observed a chimpanzee using a blade of grass as a tool a capability formerly thought to be unique to humans.

_____

Sentence Gates: the Semicolon and the Colon

Why do sentences get larger and longer as students progress from elementary school to high school to college? They lengthen because the thought process becomes more diligent. Children are noticeably declarative: "I want a popsicle." Adults are noticeably more thorough, more *causal* to be precise: "I would prefer a Volvo because it is the safest of the more reliable cars on the road in this age of eight-lane freeways, higher speeds, and increased traffic volume."

Commas assist the growing complexity of thought by setting off all of the parts within the logic of a sentence, whereas the semicolon and the colon tend to *extend* the logic of the sentence. You will notice in the following rules that these marks show up at the end of a sentence and open up the possibility of keeping the logic going. Commas usually appear just about anywhere *before* the end of a sentence. Commas define, or mark off, parts that make the internal sentence logic precise. The semicolon and the colon tend to have the larger function of saying, "This is explained by . . ." or "because . . . ."

- The comma usually signals logical detailing.
- The semicolon and colon signal logical progressions.

Notice that I have not referred here to the parts of a sentence in terms of clauses, phrases, and words. You may find it easier to perceive the unit of the sentence simply as a complete logic structure, which it is. Then the role of the semicolon or the colon is easier to see. The semicolon and the colon simply act as gates or logic links for holding sentences together.

The Semicolon

- Use a semicolon to join two logically related sentences.

> I had a bad day from the start; the car battery was dead.

> The chemist Thomas Midgley had a strong influence on the automobile industry; he discovered that tetraethyl lead could be added to gasoline to stop engine knock.

Notice the causal connection between either pair of sentences (independent clauses). The logical link is so obvious that I could have used the word *because* rather than a semicolon. Notice also that the second of the two sentences is no longer capitalized once it joins the first.

Until you become familiar with this punctuation tool, you will be prone to a common error. Look at the following sentence:

Ø   Engine knock is caused by excessive volatility in gasoline; which is inhibited by adding tetraethyl lead.

In this case, the second sentence is not a complete sentence (an independent clause). To avoid this error, do not use a semicolon until you have constructed both sentences separately. Most writers recognize complete sentences, so rely on that skill first. Then connect the sentences you want to connect. Write first. Punctuate second.

- Use a semicolon and a logic junction, usually the words *however* and *therefore,* to join two logically related sentences.

Magnesium is a very lightweight metal; **however,** it is usually alloyed with aluminum, zinc, or manganese to produce slightly heavier alloys for the aviation industry.

Writers use two groups of connectors to structure sentence logic. The simple group of logic links consists of the daily tools you use hundreds of times a day: *and, or, but, for, nor, yet, so.* These short connectors draw attention to endless logical unions, although people seldom realize how important they are. Notice the clear logic processes of these basic units.

A *and* B	I *nor* J
C *or* D	K *yet* L
E *but* F	M *so* N
G *for* H	

These conjunctions function with logical precision:

A is true, and B is true.

I call the simple conjunctions "short connectors" because they evolved as monosyllables (one sound), perhaps because everyone makes heavy use of them in speaking. There is another, larger group of connectors characterized by their length. I call them the "big connectors" or "logic junctions." They are not daily fare, but they add great precision to our logic. The dominant two serve obvious logic functions.

A *therefore* B

C *however* D

In meaning, the simple conjunctions are logically equivalent to several of the larger terms:

but = however

yet = however

so = therefore

Writers use the logic junctions (called *conjunctive adverbs*) to extend the logical relationships they can construct with a sentence.

C is true; however, D is also true

As a rule, these links simply tend to join another sentence to the parent sentence or original sentence. The construction is much the same as the one that is built with simple conjunctions.

> Pasteur could not find the microorganism that caused rabies; **therefore,** he theorized quite correctly that the cause was much too small to be seen in the microscope.

In sum, you can use the semicolon by itself or with a *logic junction* (a conjunctive adverb). If you use a conjunctive adverb to join sentences, put a comma after it because the unit then comes in three parts:

; however,

There are a number of these larger logic junctions that you will see in scientific papers. Here are a few of the more popular ones. They are indispensable in technical work.

; accordingly,

; consequently,

; finally,

; meanwhile,

; moreover,

; nevertheless,

; nonetheless,

; otherwise,

; similarly,

; subsequently,

In general, the logic junctions are identifiably quite formal and many syllables longer than the workaday conjunctions. A few other logic links you are likely to use include *then, thus, still, now,* and *also.* These are monosyllables, but they are preceded by semicolons also.

- Use a semicolon to link two sentences that you have joined with a transitional phrase.

Yet another group of logic junctions—transitional phrases—are also used to extend sentences. You could construct a variety of such expressions, but writers tend to use a dominant group that includes the following:

; in addition,

; on the contrary,

; in fact,

; for instance,

; on the other hand,

; in other words,

; as a result,

These phrases extend sentences quite conveniently and take the semicolon and a comma.

- Use semicolons to clearly separate the elements of a series within a series.

If you try to render the following structure (the one on the left with the brackets) in the simple linear pattern of a sentence, the going can be rough. The punctuated pattern appears on the right.

```
A ⎫                    A,
B ⎬                    B,
C ⎭                    C
                                    ;
D ⎫                    D,
E ⎬  ⎫                 E,
F ⎭  ⎬                 F
     ⎬                              ;
G ⎫  ⎭                 G,
H ⎬                    H,
I ⎭                    I
```

To shift from the convenient bracketed format to the written equivalent, you must divide the lesser series with commas and join them as three elements in a greater series by using semicolons. For example,

> Astronomy involves solar system studies of the sun, the planets, and the moons; stellar studies of the red giants, the blue dwarfs, and supernovas; and studies of star variations, such as white dwarfs, pulsars, quasars, and other phenomena.

In technical writing applications, the technique of developing a vertical list of some kind will often prove to be easier and more practical than a sentence for dealing with a series in a series. Engineers and scientists prefer a list; it is less complicated.

The largest ecosystems are the world's major life zones:

I. Biomes conditioned by latitude

- Tundra (boggy plains)
- Taiga (evergreen forests)
- Temperate Forests

    —Rain Forests

    —Deciduous Forests

- Grasslands

II. Biomes conditioned by precipitation

- Deserts

- Savannah

- Tropical Rain Forests

The Colon

I suspect that writers seldom use the semicolon because they are afraid that it somehow involves a lot of complicated grammar, which it does not. The colon is a different matter. In my experience, authors may simply be unfamiliar with any use for the colon other than the "list rule," and they often misunderstand that rule in any case. There are four different uses of the colon that are simple and practical tools for a writer's trade.

Because I am taking the approach that punctuation helps control the logic functions of sentence structures, observe that all the following rules for colons have a unique mathematical function: they all mean *equality*. All the rules indicate a situation in which A equals B; thus, the symbol for the colon could just as well be an equal sign borrowed from math.

- The series. Use a colon to introduce a series introduced by a sentence.

   The architect provided the prospective client with three sets of pictorial drawings:
   isotonic views, oblique views, and representational views.

The most common error results from using a colon in a sentence that did not need one.

   Ø   The endangered birds in the jungle canopy were: green, yellow, or brown.

Yes, you use a colon to introduce a series, but *only* if a sentence precedes the colon. In the preceding sentence you can see that the sentence keeps going because the verb, a linking verb, has not completed its logic. I can use the existing sentence by either leaving out the colon (you do not *always* use a colon to indicate a series) or by completing the sentence before the colon.

   The endangered birds in the jungle were disguised by their colors: greens, yellows
   and browns.

If you are not sure, you can always use the expression "as follows" in technical work. The result is usually a properly handled colon.

   The orders were as follows: look sharp, be on time, get the job done.

   The early structural theories behind bridge building depended on the following:
   stress, strain, virtual work, elasticity, and load failure.

Do *not* use a colon just because you have indented a list or a quote. This is commonly done, but you are simply seeing the most frequent punctuation error that occurs among well-educated writers. There *must* be a sentence before the colon or else you do not use it, regardless of the indented material.

Before I go on, quickly notice the "equal sign" value of the colon in the correct sample concerning the endangered birds.

> their colors = greens, yellows, and browns

- The echo effect. Use a colon to connect two sentences that say the same thing.

The "echo rule" is based on the well-known learning tool of repetition: say it twice. If two sentences essentially say the same thing, join them with a colon, creating an echo effect. The challenge is to make the repetition distinct and somewhat different from the initial comment. The usual technique is to make one sentence longer than the other one. The shorter one can be the first or the second.

short first sentence: amplification.

long first sentence: condensed version.

In the following examples, you can see that the echo rule is a conspicuous example of the equality of A and B.

> A seed is born to its fate: within each and every living organism there exists the essence of death in the gift of life.

> The coelacanth was thought to be extinct for 70 million years because it shares the attributes of very early fish species: this ancient fish is a living fossil.

Be sure that you have a sentence both before and after the colon because the colon is not signaling a list in this case. Here, the colon is joining similar sentences. Write the sentences as separate constructions first. Do *not* capitalize the second sentence after the colon.

A very practical location for either echo construction is the introduction or conclusion of a document. Using repetition in the opening remarks helps identify key points for a reader. Using the same device in a conclusion helps isolate the most important points you want to be sure a reader has understood from the discussion:

> Finally, the economic failures of communism resulted from poorly managed national centralization, and this was apparent in the fiction of socialist humanism in which rule by the masses was practiced as rule by the bureaucracy: communism was its own tyranny.

- Emphatic constructions. Use a colon to highlight the logical point of a sentence.

> Refrigeration may be a mechanical process but it depends on the special properties of one chlorofluorocarbon: **Freon.**

> George Eastman could not have developed the Kodak camera if another inventor, H. M. Reichenbach, had not invented the most important feature of Kodak technology: **celluloid roll film.**

> Robinson changed the history of medicine when he synthesized one particular complex alkaloid: **morphine.**

Notice the twist in the logic of each sentence. Something is missing: the answer! This device is a perfect way to highlight a word or phrase in technical writing in which writers are not supposed to be dramatic. The sentence modestly constructs a punchline. As always, there must be a complete sentence before the colon.

> We have one priority in the effort to salvage this ship at a depth of 300 feet: **safety.**

Notice once again that the use of a colon in this instance (called, by the way, an appositional construction) establishes a condition of equality between A and B.

chlorofluorocarbon = Freon

important feature   = celluloid roll films

alkaloid   = morphine

priority   = safety

- Titles. Use a colon to separate two titles on a manuscript.

The Couch Cadaver:
The Deadening Effects of Television

Energy Barons: The Game of Power Profits

The practice of using more than one title for a document is an old one. Formerly, titles used to involve phrases that are now passé because they resulted in very long, windy titles.

Salmonoids,

Being the Compleat Study of . . .

dahdah, dahdah . . .

The point of using the more modern two-title method is to build interest or catch the eye with the first and get down to business with the second.

Salmon Migration Controls:

An Analysis of the Failures of the Fish Ramp Technology Utilized in

Dam Construction on the Columbia River

Here, too, you see yet another colon rule designed to indicate equality of halves: A equals B.

Salmon Migration Controls = Fish Ramp Technology

If a magazine publishes such an article as this one concerning salmon, the layout designer is likely to use two different fonts, two different sizes, and maybe two different colors for the titles. Avoid these temptations in a manuscript. Place one title above the other and place a colon at the end of the first. Also, if you wish to use the double-title device, avoid the popular journalistic tool of the pseudoquestion:

Helium Neon Lasers:

Safe or Dangerous?

The second title is not a question and cannot be constructed with a question mark unless it is interrogative.

Helium Neon Lasers:

Are These New Lasers Safe or Dangerous?

A Final Note

How often should you use semicolons and colons? Judge their use by the level of knowledge you expect from your readership. Remember that these devices join the logical structures you are developing. This may be a perfectly desirable approach to making a reader see relationships in scientific and technological discussions. On the other hand, simplicity suggests an equally effective tactic: at times, and for certain audiences, relatively short sentence structures are a more effective tool, whether it is because the subject is hard to handle or because the readers might have difficulty with it.

Exercises Chapter 3, Section C

Add the necessary punctuation to the following sentences.

1. The silicon chip is the heart of the computer technology of the last twenty years the chip was independently developed by Jack Kilby of Texas Instruments (1958) and Robert Noyce of Fairchild Semiconductor (1959).

2. FORTRAN was the first programming language to gain wide success however BASIC was subsequently developed by Kemeny and Kurtz at Dartmouth University and it became the most popular language for personal computers.

3. Most forms of bacteria are harmless in fact certain bacteria types such as the Escherichia coli in the human intestines are important contributors to the digestion process.

4. Hiram Maxim developed the first fully automatic machine gun in the 1880s however a much lighter machine gun—the tommy gun—was developed by John Thompson and patented in 1920.

5. Joseph Schick invented the modern electric razor and introduced it in 1928 in addition Walter Diemer invented bubblegum in the same year.

6. *The body uses three sources of energy fats proteins and carbohydrates.*

---

7. *The months of the calendar were usually based on the names of Roman gods Roman leaders and Roman numbers the months of January March and May are derived from the gods Janus Mars and Maia the months of July and August were based on the names of the emperors Julius Caesar and Augustus Caesar the months of September October November and December are taken from the Latin words septem (seven) octo (eight) novem (nine) and decem (ten).*

---

8. *There are a number of pollutants that are commonly found in the home asbestos carbon monoxide formaldehyde lead mercury and nitrogen dioxide.*

---

9. *COBOL remains a popular computer language because it serves business and commercial applications in which the handling of commonplace arithmetic operations occurs in huge files COBOL remains an excellent choice for business needs.*

---

10. *The most common contagious disease in the world is coryza, which occurs in hundreds of strains and is readily spread by body contact and by airborne particulates carrying this dreaded annoyance this pest is the common cold.*

---

11. *The American astronomer Edwin Hubble was the first to suggest the currently accepted theory of the origin of the universe the big bang.*

---

12. Semiconductor technology led to a basic design value that is critical to the age of the computer miniaturization.

_____

13.   *The Inventions of Death*
         Weapons Patents of the Twentieth Century

_____

14.   *Maggots and Mold*
    Biogenesis and the Theory of the Origin of Life Forms
    in the Writings of Francisco Redi and Lazzaro Spallanzani

_____

Punctuate the following complicated sentence. Then, since it remains somewhat difficult to read, redesign the layout and present the sentence in an outline style, perhaps using bullets (•).

_____

Many old calendars survive and are in use throughout the world the Gregorian calendar was designed to place Easter on the vernal equinox (the first day of spring) and is based on a 365-day calendar 12 months of mixed 30- and 31-day units and a 28-day month in February the Japanese calendar is structured on a 360-day year and consists of 12-month cycles and a 52-week year but the years are numbered in epochs based on the reigns of emperors the Coptic calendar is used in areas of Egypt and Ethiopia and is based on a 365-day calendar 12 months of 30 days each and a 5-day unit to balance out the year.

_____

Other Tools

The Apostrophe

- Use an apostrophe to indicate multiples of numbers or other units if desired. Note that some authorities omit the apostrophe.

6's

***X*'s**

- Use an apostrophe to indicate the missing letters in a contraction, but avoid contractions in formal writing unless you are constructing a question that would otherwise be very clumsy.

> **Shouldn't** the Food and Drug Administration regulate the 5000 chemicals used to make perfumes and colognes?

The only contraction many writers use regularly is

let's

because *let us* sounds awkward (see also p. 136).

Possession

Writers add an *s* at the end of words as a symbol to show multiples (three dogs) or possession (Moe's tuggy). The apostrophe simply tells the reader that the writer is talking about possession and not plural numbers. To form the possessive of a plural that ends in *s*, add the apostrophe *after* the *s*.

> All three dogs' tuggies were in the box.

The facts are easily kept straight.

's	**=**	**one**	**=**	**Susan's printer**
s'	**=**	**two or more**	**=**	**the members' folders**

There are four areas of confusion regarding the apostrophe:

- Remember that the *'s* construction has to do with the owner or the owners who own. The number of items that are owned have nothing to do with the concept.

- Do not add an apostrophe to possessive pronouns. They already imply possession. If I add an *s* to a noun—*bees*—it means lots of bees. I have to add the apostrophe if I want to say *bees'* nest. But if I say *his* dog, *her* cat, or *their* car there is no need for the apostrophe because these words exist *only* as possessives.

You would not, therefore, construct words such as

Ø	his'	Ø	our's
Ø	her's	Ø	your's
Ø	it's	Ø	their's

Any use of the apostrophe with these possessive pronouns will be incorrect.

- There are a few quirky situations.

> I like Bob and Beth's new work station.

Only the last name takes the apostrophe in group ownership.

> My father-in-law's portable surveyor is a Magellan GPSNav5000Pro.

- The variations based on "one" take an apostrophe since such words are referring to someone in possession but not a specific person.

everyone's homepage numbers

one's calculator

someone's labcoat

no one's disk

anyone's opinion

The Slash

- Do not use the slash. It has little or no meaning in writing, and the constructions that can be fabricated with it are seldom spoken. If you could do a search on any one of the *Wordworks* books for slash constructions, you might find one in fifty pages of running text, although you will see a number of slashes in the subheadings of many books, including grammar books. There are very few popularly accepted slash constructions.

and/or

he/she

An author has to decide whether these constructions should be used. I generally avoid them.

The Dash

The dash is an excellent device for adding clarity *if* you type it correctly. *It is not a hyphen. It is composed of two typed hyphens.* The idea is to be quite visible in using the dash.

The dash is a way to create emphasis that allows you to avoid exclamations, underlining, and all caps. Remember, in formal writing there are

<u>NO GIMMICKS!</u>

Avoid all three of the devices used in this sample, except in headings.

Use the dash in place of a comma or commas to add emphasis or to denote a sudden break in sentence structure.

> He was a part-time program designer—and a very good one.

A dash construction can also be placed in the middle of a sentence.

> Your five grades—all of which are critical—will determine your term average.

> You have ten grades in total—two of which will not be counted.

The Hyphen

The hyphen joins words, which is a small job for a small symbol. The dash separates parts of sentences, which is a bigger job for a bigger symbol.

> She was a **part-time** employee.

> She bought a **top-of-the-line** computer—but it would not work.

Use hyphens to join words in a phrase used as an adjectives.

> It was a **one-of-a-kind** buy.

The hyphens you should see in expressions such as

> a once-in-a-lifetime opportunity

are often left out, particularly in the advertising media where most of them originate.

- Use a hyphen to divide a word (between syllables) at the end of a typed line. A far more attractive practice is to program your word processor to justify your right margin. The end-of-line hyphens will then be removed entirely.

Parentheses

- Dashes highlight. Parentheses suggest the opposite: clarifications, additional information, or afterthoughts.

  > PVC (**polyvinylchloride**) is the thermoplastic compound we know as *vinyl*.

  > The carrier of the gene code is DNA (**deoxyribonucleic acid**).

(See also the use of the comma for apposition on p. 61.)

Parentheses can be used to insert information that is awkward but unavoidable.

> Synthetic aniline dyes (**benzene was the principal raw material for the first synthetic dyes**) quickly changed the textile industry.

> Roebling's method of spinning steel cables (**he was the chief engineer for the construction of the Brooklyn Bridge**) is still used in the construction of large suspension bridges.

- Use parentheses in technical documents when you insert a numbered list in running text.

  > The survey team had to (1) determine the elevation, (2) calculate the distances, and (3) approximate the location of the structure.

- Although there is no rule that says that the popular practice of using a single parenthesis is incorrect, it is easy to make the device conform to standards.

  > 1) The initial task involved disassembly.

Add the missing parenthesis.

> (2) The next procedure involved testing.

In an elaborate outline that consists of many subsections, both the single and the double forms can be used.

Distinctive Treatment of Terminology

Italics

Originally used to identify foreign words likely to be unfamiliar to the reader, italics are particularly helpful in identifying technical jargon, particularly in the fields of data processing, computer engineering, and computer programming, in which a great many terms can be confusing. A physicist who discovers a new subatomic particle will create a new word for the quark or gluon. It is a long-standing tradition to create new words or taxonomic designations regardless of the scientific field. The glaring exceptions to this practice are data processing (DP), programming in general, and all the computer-related disciplines. Here, there is a strong preference for reusing standard English vocabulary—with entirely different meanings, of course.

DP languages indicate the problem.

> Another error that does not cause the program to abort with a run-time message occurs when the programmer leaves out the & (address of) operator on the variables in the call to scanf. Because scanf does not know where to find the variable name, it is unable to store in them the values entered by the user. In this instance the program runs toward completion using whatever garbage values were originally in the memory locations.

In general, computer technology demonstrates a similar style of terminology.

> After the ROM-BIOS takes care of the shift and toggle keys, it needs to check for some special key combinations, such as the [CTRL]-[ALT]-[DEL] combination that reboots the computer. Another special combination is [CTRL]-[NUMLOCK], which makes the computer pause. Finally, if a key action passes through all that special handling, it means that the key is an ordinary one.

Internet vocabulary has evolved in the same way.

> Most anchors are in the form of <A HREF="URL">, where URL is the URL of what you are pointing to. HREF stands for "Hypertext REFerence." For example, the NSCC Web server is at "http:// nscux.sccd.ctc.edu/" A sentence that contained a link to that address would look something like Figure 13. The words between the open and close of the anchor ("NSCC Web server") would be displayed as a hyperlink. Selecting that link within the Web browser would cause the browser to load the NSCC Web home page.

Technical jargon can simultaneously build exacting commentary and confuse the uninitiated. This is always a risk writers take in their work, but computer-related terminology deserves special attention. Whereas most disciplines have developed unique vocabularies, the computer-related technologies recycle the English language. The words *transistor* or *bacillus* could never be confused with other words. On the other hand, *import, file map, move file, pruning directories, disk compressing,* and a thousand such terms pose a special problem because readers cannot always immediately identify the discrete term or its function, and they end up searching through many mental definitions of the same words. Italics can help clarify this situation. (Underlining text is the way to indicate italics on a typewriter.)

Computer language assumes multiple roles also. Input is a noun. Input is a verb. Input is an adjective.

> I have the input.
>
> I input the data.
>
> This is the input folder.

This problem is unique to the computer fields. *Bacillus* (in the field of biology) is not only a unique word; it is always a noun and not a verb or adjective. Here are suggestions for managing computer language problems.

- Use italics to highlight words that are critical to understanding a text, particularly if the vocabulary is likely to be new to the reader. Usually the highlighted terms will appear in a glossary.

> When you are coding, you must pay attention to *logical errors* because the C++ *compiler* will not catch those errors for you. Make sure you use the right mathematical formulas in your calculations. If you use a *loop,* make sure that you go through the loop to be sure you do the three required actions of a loop: *initialize, test,* and *update.*

- In general, use italics (or underlining, if you are typing) to define or introduce data processing or computer programming terminology, particularly if the word is a reuse or variant of our daily vocabulary. The following sample would be typical of a typewritten document.

> When you turn on a computer, the first task the PC performs is a self-diagnosis called a power on self test (POST). During the POST, the computer identifies its memory, disks, keyboard, display system, and any other devices attached to it. Then the computer looks for an OS to boot. A PC looks for the OS on the primary floppy drive first; if it finds a bootable OS there, it uses that OS; otherwise, it looks on the primary hard disk.

- In DP or programming, consider using italics or underlining. The use of italics is intended to increase readability and can, therefore, be used throughout a document, particularly if the document is intended for a novice.

To use italics, select a single word or series of words by highlighting and then click *Italics* from the tool bar or menu (depending on your computer's word processing program). The newer twelve-function computer keyboards may also allow your program to perform some convenient shortcuts for setting up extensive use of italics.

There are a number of other traditional uses of italics:

- Titles and subtitles of books, periodicals, pamphlets, dissertations, theses, manuscripts, papers for meetings, and corporate reports are printed in italics when they are mentioned in another document. In general, if the item is from a set or a series, use italics. Underline the titles if you are using a typewriter.

- Frequently mentioned titles can be abbreviated but remain in italics.

> The Velikovsky text *Worlds in Collision* is very controversial.
>
> Some authorities consider *Worlds* to be pure fantasy.

(Notice that a title is singular even if it is constructed as a plural. The internal logic of the title does not affect the fact that it is a single title.)

- Use italics for a letter used as a letter.

> *X* represented the quantity.

Use italics for a word used as a word.

> *Endorphins* is a shortened form of *endogenous morphines*.

Use quotations marks for a standard term that is used in a nonstandard manners.

- Italicize foreign words, but do *not* italicize foreign place names of towns, parks, streets, structures, and so on.

> We usually stay in Montparnasse near the Jardins du Luxembourg. The restaurant on our street serves an excellent *blanquette de veau.*

- Do not italicize any letters of the Greek alphabet, even though they are foreign words.

> Alpha   Beta   Gamma   Delta   Omega

- *All* the rules for italics (or boldface) are important applications that assist the author of technical or scientific documents.

- *Underlining.* Italics is generally preferable to underlining for all the preceding applications. Perhaps the primary exception would be underlined headings in a manuscript in which underlining can highlight the heading.

Boldface

Boldface is increasingly popular and is now often seen in running text, particularly in corporate manuals and technical manuals in which boldface is often used in place of italics or as a companion method of highlighting when both systems are desired. If you have two sets of jargon, for example, or two types of text body, the use of two highlighting systems might be useful. For example, the *Writer's Handbook* generally uses italics for cited words in the running text and boldface for key words in the samples.

You will see the two used together throughout much of the *Writer's Handbook* (see, for example, p. 170).

Quotation

Although considerable attention can be dedicated to the details of correctly quoting material, for your purposes the use of quotation is secondary to the many other varieties of evidence you are likely to use in technical work. Quoting an authority is an acceptable practice in the sciences if a writer is concerned with discussions of theory or conjecture or analytical observations. Far more critical are the primary scientific findings—numerical or otherwise. Engineers and engineering technicians tend to cite statistical outcomes of water purity, charts of sonar readings, tables of infestation rates, graphs of demographic shifts, photographs of chromosome changes, and other types of evidence. However, if you use a quotation, a few simple rules can make quotations quite manageable:

- What are popularly called *quotation marks* are the double quotes used here. Identify the speaker with a quotation tag (*he said, she said*),

   He said, "I will analyze the data."

- The tag is an important feature of the quote. It indicates to the reader that a quotation is about to begin, but it is usually also used to identify the author of the quote. The author is commonly identified by his or her career, and the source of the quote is often identified as well.

   The German physicist Max Planck commented, (followed by a quote).

   James Watson, well known for his work on the molecular structure of DNA, explained his ambitions in the *The Double Helix:* (followed by a quote).

- If the quote is a sentence or two, incorporate the material in the text. Use a quotation tag followed by a comma to introduce the quote. You may use a colon if the quotation consists of more than one complete sentence and the tag is a sentence.

- In general, if the quote is quite long (ten or more typed lines), indent the entire quotation, double space the material, and leave out the quotation marks.

- Use ellipses (. . .) to indicate words left out of the quotation to make the quote conveniently fit a sentence.

   NOTE: Most quotations are taken out of their context, so you do not need the ellipses to indicate that there was more material. Use it only to indicate missing material from your *specific* quotation.

As an example, I will use the following quotation:

   In science, there is only physics; all the rest is stamp collecting.

   –Ernest Rutherford

I can alter the quote if I delete sections of it.

> Ernest Rutherford was fond of arguing that "In science there is only physics. . . ."

- The following figure illustrates the rules for punctuating the end of a quotation.

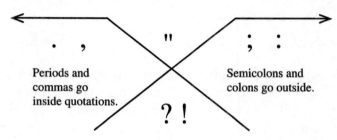

Periods and commas go inside quotations.

Semicolons and colons go outside.

The logic of the sentence will determine the position of question marks and exclamation points.

The chart indicates the following applications:

> She said, "I will go."
>
> John said, "I will go"; however, Mary did not.
>
> She asked "Will he go?"
>
> Did he say "I will go"?

It is obvious that you need to attach the question mark to the question, and you can see that the question can be inside or outside of the quotes. Writers get confused about the other punctuation because the logic to the rules is not entirely evident. British writers in fact, handle them in reverse!

- If your subject matter is data processing, computer programming, or Internet communication, avoid the common practice and do *not* put periods or commas inside a quote. Periods are gremlins in cyberspace. They may be proper in the rules of writing but they can sabotage your keyboard activities, programs, and electronic communications.

- In data processing documents and computer programming papers, I suggest that you handle your material largely in terms of the display techniques used in mathematics (see Chapter 8). Double space; indent; enter the material as an indented, single-spaced subset; and exit the display. Do not use quotation marks if you do not need them.

- Single quotation marks are seldom used. They are usually intended to identify a quote within a quote.

> She said, "I will go if he says 'I apologize.'"

- If you want to cite a word, use italics or double quotation marks. Do *not* use single quotation marks.

> A *byte* is one keystroke of memory. In practical terms, a byte is one letter of a word.

- Many writers assume that double quotation marks are useful only for full quotations, and they will then use single quotes for cited words. In fact, double quotation marks are a practical convention for citing *any* material, and the single quotation marks are uncommon.

Paraphrasing

If you borrow ideas but not the exact wording, you are paraphrasing. In this case, the use of quotation marks could be very misleading since you are not repeating the same words that someone else said. However, you want to be sure to explain that the ideas are borrowed so as not to mislead the the reader into thinking that the ideas are yours.

The solution is to tag a paraphrase and state the material in your own words.

> Biologist Philip Talmidge felt that the well-known demonstration of the polymerase chain reaction (PCR) was certainly not going to evolve into a Jurassic jungle (Smith 209).

The end of the paraphrase is often, though not always, signaled by a reference to the source, either in an author-date citation or with a superscript that refers to a footnote or an endnote.

Exercises Chapter 3, Section D

Add the necessary punctuation to the following sentences.

1. Bits are the smallest units of information on a computer they are equivalent to 0s and 1s.

2. The redwoods heights make them the worlds tallest standing trees.

3. A ponderosas bark is one reason the tree is distinct in appearance.

4. The compounds complexities made them difficult to manufacture.

5. Aspirins initial source was the bark of a tree.

6. Bardeen and Brattains combined knowledge resulted in the first transistor in December 1947.

7. After leaving Fairchild Semiconductor Moore and Noyces plans to manufacture integrated circuits resulted in the formation of the Intel Corporation.

8. *Someones discovery of a spontaneously developing film at Agfa led to Edwin Lands invention of the Polaroid Camera twenty years later.*

_____

9. *Catgut the string that is used in surgical procedures was originally taken from the intestine of sheep not cats.*

_____

10. *Nitroglycerin dynamite TNT and other blasting compounds all of which are based on nitrogen compounds found their way into the arsenals of war.*

_____

11. *Pyrex a particularly tough form of glass consists of boron oxide 12 percent and silicon oxide 80 percent. Corning a New York firm developed the product in the 1910s.*

_____

12. *The CD-ROM compact disk read-only memory was introduced in 1984.*

_____

13. *Research on the development of the CAT computerized axial tomography scan began in Johannesburg, South Africa.*

_____

14. *Many Americans who have received the Nobel prize were immigrants and these include the first American man to receive the prize the German born American physicist Albert Abraham Michelson and the first American woman to receive the prize the Czechoslovakian born biochemist Gerty T. Cori.*

_____

CHAPTER 3 SENTENCE SYMBOLS: PUNCTUATION LOGIC

15. *Because there were no semiconductors at the time a state of the art computer from the 1940s was enormous but could do little more than one of todays pocket calculators.*

_____

16. *The initialize expression is executed once, at the beginning of the loop prior to any other loop statements being executed. The next statement the test expression is then executed.*

_____

17. *Select dial up adapter and click properties and make sure TCT/IP is checked. In the next window under Manufacturers click Microsoft.*

_____

18. *Failure of the checksum test for any of these ROM modules will cause an xxxx ROM error to appear.*

_____

19. *Caution: repartioning your hard drive with Fdisk will destroy your data.*

_____

20. *Johanson explained that we have an excellent set of footprints of Australopithecus afarensis in Tanzania.*

_____

21. *Tolitson defended the use of modern nuclear power plants and told the audience, When a powerplant goes critical, a steady and stable chain reaction is created; however, he had to explain to the audience that a nuclear weapon depends on a very different uncontrolled state in which the mass becomes, in his words, supercritical.*

_____

Subjects, Verbs, and Objects

4

Grammar consists of those conventions that reflect the common practices of sentence construction. I could call the conventions "rules," but they are more on the order of observations. The sentence is the base unit of thought for speaking and writing. I have suggested elsewhere (in *Basic Composition Skills*) that the structure of thought is much more fragmented than that of speaking or writing and that fragmentation is one reason the sentence fragment is a frequent element in our speech patterns. In writing, of course, what is called the *complete sentence* is the base unit, and fragments are considered to be malfunctions.

Sentence Logic
Sentence Symbols
SENTENCE CONVENTIONS
Word Rules
Number Rules

Traditionally, the sentence has been defined as a group of words that express a complete thought. This is a perfectly acceptable definition, but it generates the question, What is a *complete thought?*

Thought is the more complicated of the two words and is, in fact, the subject of entire scientific disciplines. I prefer to try another approach, based on the simple but reasonable observation that a sentence is usually about a thing in motion. A sentence, at minimum, must have a subject and a verb (or activity):

> I run.

> You run.

The sentence—any sentence— simply reflects the animation in the world around us. Science fiction writers have pondered over silicate life forms, but I do not know what would become of a science fiction novel based on a crystal culture, because crystals do not *do* anything. At least they do not do anything in a time frame that has meaning to us. Animation is a fundamental principle for you and me. We understand the world only in terms of performance. There is even a verb that is used to animate a steady state, to indicate only the presence of a thing: *to be.* If I say, "She is in the coffee lounge," I use a verb to give her a reality.

Everything in our world runs in terms of time. Nothing stands very still. Time *is* motion. A sentence is a law of physics in some sense, and, indeed, it will collapse if I defy the thing-in-motion principle. Observe what happens when a basic sentence is constructed or deconstructed.

Subject	Verb	Objective*
I	sent	a letter
(omit)	sent	a letter
I	(omit)	a letter
I	sent	(omit)

Try to read the constructions from left to right. If you make a fragment out of the structure, the logic breaks down. The effort to reach absolute zero (zero degrees Kelvin) is an effort to see whether motion can be stopped—because, if it can, then time ceases. But can you imagine matter in the absence of time? Can you imagine motion by itself in the absence of something moving? Can you understand the activity of something that has no motion? No, these conditions, like the sentence fragment, have no logical relationship to our animated world.

You use this basic subject-verb or subject-verb-objective relationship in virtually every sentence you construct. You then make each occurrence unique to the circumstance it describes. In other words, every sentence becomes a new logical observation in which you set up a number of conditions to explain new situations. Each of these sentences indicates an outcome with a verb that animates some process in the passage of time. You and I move from one of these instances to the next in each sentence of each hour of each day of each year of our lives. For example, look at these variations on a theme:

I run.

In order to stay healthy, I run.

In order to stay healthy, I run at least three miles three times a week.

Here you can see that the second structure adds conditions to the first. The third structure adds still more. A reader then examines the "if-then" relationship to perceive my meaning and to decide whether it is true. Then the reader moves on.

There are, then, three primary elements of the sentence structure:

- The simple subject

    A local college **instructor** was granted the patent in 1996.

---

* Sentences do not communicate random behavior. There is always a scheme to the human imagination. We try to keep entropy at bay, so actions have "objectives," what is ordinarily called the **object** of the verb. The objective is the goal or mission of the verb. This is most obvious in the large family of verb actions called **transitives**. The transitive verb lacks logical coherence without an objective.

- The simple predicate (verb)

    The fermion **represents** state-of-the-art research in particle physics.

- The simple objective

    The research examined the **data.**

As I add conditions to the basic units, the structure becomes more complex. I then look at groups of words—usually directly surrounding the basic subject, the basic verb, or the basic objective—and I then provide the details specific to these units.

- The complete subject (subject and details)

    The brilliant maverick Linus Pauling, recipient of two Nobel Prizes, proposed the helix structural concept before Crick and Watson pioneered the double helix theory.

- Complete predicates (verb *and* details)

    The free-running cutthroat salmon **return to the streams of their origins at a certain stage of the reproduction cycle of the species.**

The predicate is often complex because the action of the verb is usually the focus of interest. There are patterns to the way the action is explained in a sentence. The verbs themselves determine the pattern. *Intransitive* verbs call for no further explanation.

    The motor died.

    We laughed.

Transitive verbs complete their logic with some kind of objective or explanation.

    I slammed . . . .

    I slammed the door.

    We distributed . . . .

    We distributed the materials.

Transitive verbs have to be explained; they have meaning in relation to an objective (the object). There are two types of objectives:

- **The direct object** These nouns identify a person or thing that is receiving the action of the verb. Many verbs (the *transitives*), make little or no sense without an *object* to complete the idea—or an *objective* for the verb action, to put it another way.

He closed the **door.**

She started the **car.**

- **The indirect object** These nouns or pronouns indirectly receive the action of transitive verbs. They appear only in sentences where the verbs are a *giving* action.

    The lab technician gave **her** the report.

    The director awarded **him** a prize.

    The officer lent the **chief** his file.

The *linking verbs* form a third group: *am, is, was, were,* and so on. They link the subject with its *complement,* which is a word or phrase that describes or names a subject.

    She **is** the mayor.

    They **are** intelligent.

    The strongest one **was** Beth.

The logic of linking verbs usually calls for a completed linkage (called a *subject complement*). The situation is quite similar to that of transitive verbs that need an explanation.

    I am . . . .

    I am lucky.

If linking verbs are used to construct tenses for other verbs they simply support the other two categories of verbs.

    We **were** distributing . . . .

    I **am** working.

The subject complement consists of whatever details describe the subject. The verb links the subject and its complement. There are two different types of subject complements:

- **The predicate nominative** The predicate nominative (a noun) renames the subject by way of a linking verb, usually a form of the verb *to be*.

    Cats are **felines.**

    He was a **leader** without knowing it.

    She was a **chemist** at Dupont.

- **The predicate adjective** The predicate adjective follows a linking verb (often a "to be" verb) and describes the subject.

> The horizon is **orange** today as a result of pollution.

> The marathon runner was overly **fatigued** because of the elevation.

These primary patterns are fundamental to the way in which sentences can be constructed. Depending on whether the verb type is intransitive or transitive or linking, you will repeat one of the three basic patterns in every sentence.

subject	$\rightarrow$	verb		
subject	$\rightarrow$	verb	$\rightarrow$	objective
subject	$\rightarrow$	verb	$\rightarrow$	complement

There is no sentence without something (subject) and its action or state (verb). If the action calls for completion, the objective of the action is named, or a complement is added to complete the logic of a linking verb by either renaming or describing the subject.

The performer is the subject—usually. These front-loaded subjects identify themselves before the action starts. This structure appears to be fixed but is, in fact, flexible. This flexibility makes the language foreign to a computer and the computer foreign to the language. You can reverse what appears to be the fixed pattern of the sentence and create an inverted sentence.

He started the machine.

subject	$\rightarrow$	verb	$\rightarrow$	objective	(the usual pattern)

The machine was started by him.

objective subject	$\rightarrow$	verb	$\rightarrow$	subject	(the inverted pattern)

In the second sentence, the objective becomes the passive subject. I use the name "objective subject" in this situation to point out that the flexibility of English is nowhere more apparent than in the ability to reverse a sentence and construct it more or less backward (although the result is often criticized).

Consider the following example:

> He completed all the calculations.

subject	**verb**	**object**

It is just as easy to front-load the *object,* which then becomes the subject of the verb.

> All the calculations were completed by the student.
>
> **objective subject**   **verb**

If I use the same sentence logic for both constructions you will see the shifting that can occur.

> He wrote the report.
>
> The report was written by him.

This construction is referred to as *passive* because the subject is not the actor acting out the verb. If, however, the objective of the verb is considered more important than the subject, then the passive structure is a valuable alternative. In technical writing, commonly the objective of an action is important, and a writer often wants to focus the reader's attention on the verb and the objective. The question is when and how often to sacrifice the subject (see also pp. 117–119 and pp. 178–179).

The passive allows the subject to be moved to the end of the sentence, in order to receive contrastive emphasis

> The contest wasn't won by the boys, but by the girls.

The passive allows the writer to avoid mentioning a specific subject (by omitting the *by-*phrase), and focus rather on the activity taking place:

> The thief was caught within 1 minute.

However, I noted earlier, many people hold the idea that frequent use of the passive construction is bad style.

For a discussion of verb errors see pp. 169–179.

Exercises Chapter 4, Section A

1. Identify the direct object in the following sentences.

_____

The British "Colossus" computer contained 1500 vacuum tubes.

_____

In 1943 Colossus cracked the German Enigma codes.

_____

Developed long before the war (1919), the Enigma code machine could handle 22 million code combinations.

_____

2. Identify the indirect object in the following sentences.

_____

Kodak manufactured inexpensive cameras for consumers.

_____

In those days, Leica and the other European companies did not want to produce inexpensive cameras for the mass market.

_____

Kodak gave every buyer an affordable product.

_____

Later, in 1939, the Kodak Super 620 provided photographers an exciting innovation: the first automatic-exposure system.

_____

3. *Identify the* predicate nominative *and the* predicate adjective *in the following sentences.*

The airplane recording system called the "black box" is actually bright orange.

The first coast-to-coast highway was the Lincoln Highway (much of which is Route 30), and it was completed in 1923.

---

4. *Change the following passive constructions to active constructions.*

Major American dams are so large and costly that they have to be funded by the federal government.

An endless stream of labor-saving devices—from irons to microwaves to power mowers—have been designed by inventors.

In modern wars, 50,000 rounds of ammunition are fired by soldiers for every soldier who is killed.

It is estimated that 20 million megawatts of electricity could be generated by wind-powered generators.

Sentences, Clauses, and Phrases

Each chapter has opened with an illustration that indicates the level of discussion in terms of a simple divisible pattern. There is a similar pattern or hierarchy for the fundamental principles of sentence organization.

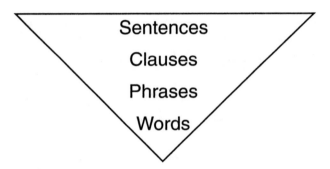

From greater units to lesser units, the pattern of our language has distinct and observable characteristics. The sentence is, of course, the fundamental unit of our logical thought processes. Clauses, phrases, and words shape and define the sentence. The smaller units generally fit within the larger units, but size is only a convenient perception since a sentence takes only two words to construct, and some exist as single words (Halt!, Speak!).

Sentences

The sentence is the basic unit of the language process. People use sentences, or variations of them, to construct logical observations about their world—always, of course, in terms of that primary assumption that something must *do* (or *be*) something. There must be a subject and a verb, and the concept evolving from the action usually must be resolved. We "complete the thought," to use conventional wisdom.

There are four basic groups of sentences (see also p. 27):

- **Simple sentences** contain one independent clause and no dependent clauses.

    I worked alone that night.

- **Compound sentences** contain two or more independent clauses and no dependent clauses.

> I worked alone that night, but there were other people in the building.

- **Complex sentences** contain an independent clause and one or more dependent clauses.

> I worked alone that night after I had spent the day in the field.

- **Compound-complex sentences** contain two or more independent clauses and one or more dependent clauses.

> I worked alone that night; however, there were other employees upstairs, which is
> why I didn't worry about security.

At the outset of the discussion I pointed out that the additional developments within the elements of a sentence establish the logical conditions of the subject-verb relationship.

> A wilting tomato plant will quickly die without water.

I take the basic logical understanding that living things die and surround the specific application of the logic with words and phrases that condition the observation for a specific set of circumstances. Any words that are added to the subject and verb add a wide variety of descriptive conditions to make the sentence logic precise. In other words, the sentence usually grows beyond the basic elements of the thing and the action (or the thing and its complement if the logic involves linking verbs). I usually want to qualify or clarify both the subject and the action. For every sentence that states a simple two-word declaration, there can be a thousand longer versions to explain it.

> Plants die.

Language development is this odyssey from the general to the specific.

> A wilting tomato plant will quickly die without water, and hybridized varieties are
> particularly prone to the problem.

As a second step in the order of events, I usually explore or explain the rationale of the observation. In other words, I try to explain *why*. It is this causal connection that leads the simple sentence into the worlds of the compound and complex structures that are constructed by joining sentences.

> A wilting tomato plant will quickly die without water **because** the maturing fruit ex-
> hausts the existing water supply in a matter of hours, which is a particularly common
> problem among hybridized varieties.

Clauses

A group of words with both a subject and a verb constitute a *clause*. They are of two types.

The Independent Clause

In effect the independent clause is what I call a "sentence." It is a construction that contains a subject and a verb, and it is independently complete as a logical statement. It cannot begin with a subordinating conjunction (see p. 127).

> The geology of the basin dates from a very early era.
>
> The Aleutian chain was once a land mass connecting Siberia and Alaska.

I can coordinate, or join, these clauses (link the sentences together) with coordinating conjunctions, semicolons, or conjunctive adverbs.

> The first jet engine patent was issued to a British engineer in 1930, **but** it was a decade before the engine was used for flight.
>
> A number of inventors proposed cambered wings for gliding; the Wrights demonstrated the same lift but with a lift force created by the first airplane engine.
>
> John Colt perfected the rotating barrel technology for his "revolver"; **however,** interchangeable parts and mass production were the real innovations of his company.

The Dependent Clause

A clause can have a subject and verb and still not make a sentence. In this case, the thought is incomplete. As I prefer to say, the animation in such a clause is not completely explained because it is only half of a causal observation. Dependent clauses depend on other clauses for meaning, and so they are said to be "subordinate" clauses. They are joined with a subordinating conjunction to the parent sentence to enlarge or complete the original logic structure. They do not stand alone.

> Ø   Computers are built with three types of components. Because they must have processors, memories, and I/O devices.

You can see that the second clause has a verb and a subject—but the logic usually *depends* on the first clause, and so they must be linked. In terms specific to grammar, the clause indicates its dependency by beginning with a subordinating conjunction, and thus it must be linked to the first clause. (See the discussion of subordinating conjunctions on p. 127.)

She went back to the lab **because** the meeting was not very interesting to her.

The 68000 was unique **because** the chip was a hybrid between 16- and 32-bit architectures.

Conventional subordinate clauses are of three types depending on their role as adjective, adverb, or noun.

- **Adjective clauses** function as adjectives by modifying nouns, and they usually begin with a relative pronoun: *who, whose, whom, that,* or *which.*

  Archaeology, **which began as a pastime for wealthy adventurous gentlemen,** has become a scientific discipline of great rigor.

  The floppy disks **that came with the computer** were defective.

  The discoverer, **who had not been identified,** was located in Australia.

- **Adverb clauses** function as lengthy descriptions that explain (modify) verbs.

  The competitors abandoned steam engines **after Henry Ford demonstrated the commercial success of the gasoline engine.**

  The microscope existed **before the microbial theory was proposed.**

- **Noun clauses** are entire groups of words that, taken as a whole, function as a single noun.

  Why the Three Mile Island nuclear accident occurred was not clearly explained at the time.

  What you will find in the storeroom should be adequate.

Phrases

Phrases are word clusters. They can be as large as or larger than clauses, but they lack verbs. Phrases, although there are a number of types, are often called "describers" or a similar term, which suggests that these word groupings are used to color and fill in language pictures. Every writer paints with words and phrases, but because I am exploring the writing of technicians and engineers, I see the process differently. I would say that phrases *quantify* or *qualify* or *clarify,* and I think of phrases as *quantifiers* or *qualifiers* or *clarifiers.* In other words, I use phrases to shift language from general to specific.

You may have heard that Eskimos have a great many words for *snow.* You may question whether the folklore is accurate, but you can be sure that if the Eskimos add *phrases* to describe snow, there will indeed be many hundreds of possibilities for fine-tuning the accuracy of the description. Phrases add precision to our language. Sentences usually address the condition of a *specific* situation rather then a general situation.

A dog will not chase a cat even though it is an instinctive behavioral trait.

This comment seems unlikely as a specific statement, does it not? If I qualify the comment and make it fit the specific circumstance, it will be quite precise. A phrase, as part of a sentence, is specific to the moment and makes the language precise in a given circumstance.

> **After the proper lessons,** a dog will not chase a cat even though it is an instinctive behavioral trait.

Phrases contain no action. They are collective descriptions (as opposed to single-word descriptions) that function to describe subjects or verbs. There are a number of types of phrases including, adjective phrases, adverb phrases, prepositional phrases and verbal phrases.

Prepositional Phrases

Perhaps the most common of phrases, the prepositional phrase is composed of a preposition and the object of the preposition (the thing that is being *placed* or *positioned* by the description). Modifiers are usually part of the phrase (see also p. 125).

> Surgery was often the only course of action available **during** the war.

> Elective surgery, **without** due consideration of alternatives, is not true elective surgery.

Verbal Phrases

Three subtypes of verbal phrases correspond to the verbal types (described on pp. 128–129). Each type of phrase consists of the verbal and the logical constructions shaped by it.

- **The participle phrase**

  > The dam **proposed by Dan Brockman** is still on the drawing board.

  > The team **having the fastest runners** will win the soccer match.

  > **Having finished,** he decided he could leave the examination room.

- **The gerund phrase**

  > **Treating patients with antidepressants** has become a substitute for psychiatric counseling.

  > **Developing a good resume** can be quite difficult.

- **Infinitive phrases**

  > **To understand the ASTM standard specifications for construction materials,** you must write to the American Society for Testing Materials.

To determine the fire resistance of structural elements, the structural engineer must consult the *BOCA Basic/National Building Code.*

To indicate a solution to both problems at this time would be foolish.

The explorers went to South America **to seek gold for the crown of Spain.**

Homeowners will usually consult their local building department **to look for appropriate zoning ordinances.**

Exercises Chapter 4, Section B

A. *Identify the four basic sentence types in the following samples.*

_____

1. Seawater contains between 3.3 and 3.7% salt, although the salt content of the Dead Sea is an extraordinary 25%.

_____

2. The MS Equation Editor is one of the programs that is sold with Microsoft Office 97 and Microsoft Office 2000; however, if one version of Microsoft Office is not already saved on the hard drive, the user needs to install one to access Microsoft Equation Editor.

_____

3. The moon's diameter is 2159 miles, and it is 27% the size of the earth.

_____

4. If you have one new message, wait a few seconds so the program has time to save it to the hard disk, but if you have received more than one new message, click on the **Next** arrow, which is an icon at the lower right-hand corner of the message, to display the second message.

_____

5. In its day, the Pennsylvania Turnpike was celebrated as the first modern highway, but it is no longer considered as safe as today's superhighways because engineers now use much wider roadbeds and medial barriers.

_____

6. Once you have opened all your new messages, exit the program.

B. Correct the following samples, each of which contains a dependent clause.

1. The pencil with an eraser attached to it was a patent applied for in 1858. Which was granted to Hyman L. Lipman, who deserves credit for sticking the two items together.

2. Gum rubber products were being developed by the mid-nineteenth century. Although Lipman may not have been the person who first observed that varieties of gum rubber could be used to erase.

3. To view Code Window, click the right-hand button of your mouse on Form 1 in the Properties window, but select View Code. Because then you can type codes in the View Code form.

C. Underline the prepositional and verbal phrases in the following sentences.

1. During the 1980s, there were many advances in computer technology.

2. Environmental concerns were not taken very seriously before the publication of Rachel Carson's book The Silent Spring.

3. Chemical industries, under pressure from the EPA, were supposed to reduce chlorofluorocarbon (CFC) production by 50% by the year 2000.

_____

4. Finding the time to do field research was a major problem for the archaeologist.

_____

5. To locate the source of this opal you should consult a Brazilian gemologist.

_____

6. Having timed the crocodiles on land and in water, researchers realized that the animal is considerably faster on land, but only for the short distances.

The Parts of Speech

The language system is traditionally divided into eight groups of words. The engines are the nouns and verbs. The pronouns make life a little simpler by providing handy substitutes for nouns so that there is no need to keep using the same precise word again and again: the protozoa, the protozoa, the protozoa. Pronouns are one of the three groups of words that explain nouns or verbs. Pronouns refer to (or stand in place of) nouns to avoid their tiresome repetition. The other two major groupings are the adjectives that describe nouns and the adverbs that describe verbs. These three groups can be thought of as "descriptors" or words that describe or reference other words. A speaker or writer *describes* nouns and verbs to give any sentence a specific use in a specific situation at a unique moment in time. Adjectives describe nouns and other adjectives (*five big, healthy, green* trees), and adverbs describe verbs, adjectives, and other adverbs (*usually* heard, sang *warmly, never* went, *quickly* tackled, *always* felt, was *not* frustrated). These five classes of words constitute nearly 99% of the language.

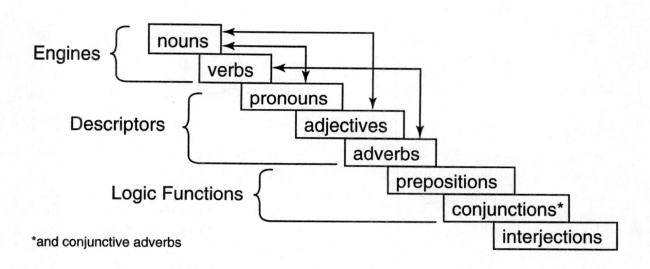

Engines { nouns / verbs

Descriptors { pronouns / adjectives / adverbs

Logic Functions { prepositions / conjunctions* / interjections

*and conjunctive adverbs

A third category is composed of what might be called the "executes" or "logic functions." It takes very few symbols in math to designate functions. The same is true with language, so the remaining 1% of the language includes a group of logical controls for the patterns you create. For example, you need to know when and where (*after, on* are examples of prepositions), and you need to know whether to add or subtract (*and, but* are examples of conjunctions), or you need to establish some more sophisticated causal links (*however, therefore* are examples of adverbial conjunctions).

Needless to say, there have to be rules to hold all the elements together. Let's look at these elements in the order presented in the diagram, which begins with the one part of speech that seems to give everybody the least difficulty.

Nouns

- Nouns identify persons, places, and things but also ideas, concepts, events, and activities.

    betrayal, democracy, flood, jogging, tennis, storm, hill

- Unlike other parts of speech, the noun has a social matrix, so specific persons, places, and things are capitalized.

    Michael Faraday, Hong Kong, the White House

- Nouns also show gender or indeterminate status.

    Ralph, man, Susan, woman, mail carrier, drafter

- A great many nouns have evolved as a result of combining two or more words. These are subject to considerable variation in spelling.

halfback	policy maker
half-dollar	mother-in-law
half sister	fellow employee

Verbs

You will notice that it takes ten times more space to discuss the verbs. I could just say they are more complicated, but perhaps I can suggest why.

Mass is visible. Energy is less apparent. Motion (which most verbs represent) brings mass and energy together in ways that explore activities and time. Mass (or a noun) is reasonably clear. Motion is more complicated because energy is more abstract and because time involves abstractions that are represented in what are called "tenses." For these reasons, perhaps, the complex life of a verb is at the heart of the language system. The verb captures a specific instance of energy.

    I work.   It boiled.   They exploded.

The verb serves to identify a physical or a mental action or a state of being. The first two types perform conspicuous actions.

The toxic waste dump **neutralized** a wide variety of petrochemical pollutants.

Phobic personalities **imagine** great risks when there may be none.

The eight pairs of birds they located **were** the only survivors of the species.

Action Verbs

There are two types of action verbs, as noted earlier. Action verbs often have no logical meaning without additional words to complete the idea of the action.

I can say, "Audubon painted."   (intransitive)

I can't say, "Audubon brushed."   (transitive)

The second instance is a verb that cannot stand alone. I must add the objective.

Audubon brushed **sienna tint into the chest coloration of the bird.**

The action verbs are usually unmistakable if they are transitive.

They **projected.** . . .

They projected the results.

The sales rep **sold** . . .

The sales rep sold him the calculator at cost.

The logic collapses if the verb idea is not completed. Other action verbs can stand alone, as noted earlier. These are the instransitive verbs.

I breathe.

You teach.

He fights.

Some verbs can be both transitive and intransitive.

She sings the blues.

She sings.

State-of-Being Verbs

The verb that is usually used to indicate "states" is *to be* and its many forms—*am, is, was, were,* and so on. They are intransitive verbs, so they do not take objects, but they must link. They are attached to predicate adjectives or predicate nominatives. Notice that the following sentences cannot stop after the verb, and these verbs are generally called *linking verbs*

The damage **was** extensive.	(predicate adjective)
A good educator **is** a good listener.	(predicate nominative)

Adverb modifiers sometimes follow verbs that are used to indicate states.

We were **nearby.**

The whale was **below the boat.**

Action verbs signal genuine activity (not states). *States* refer to conditions of being or existing but also include our sense responses.

I am . . .	We felt . . .	She was . . .
I am late.	We felt very happy.	She was sad.
Air brakes sound . . .	Soy sauce tastes . . .	
Air brakes sound shrill.	Soy sauce tastes salty.	

Other similar verbs include *to smell, to feel, to appear, to look,* and other sensory-based verbs.

Principal Parts of Verbs

In order to construct a historical point of reference for an action, the verb must be systematically altered to move it to the past or to the future or elsewhere in time. The principal parts of a verb are the *present infinitive,* the *past tense,* and the *present* and *past participles.* All the other forms of a verb can be derived from these to create time conditions or action contingencies such as *I have done* or *I will have done.*

Present Infinitive	Past Participle	Present Participle	Past Tense
to calculate	calculated	calculating	calculated
to investigate	investigated	investigating	investigated

Adding *ed* or *d* creates both of the past forms, unless the verb is irregular. The present participle is always constructed with the *ing* ending. These endings are standard practice for English speakers, and they reflect historical practices. As I noted earlier, grammar is largely composed of observations or facts and not "rules"; people speak by convention or by common agreement. Otherwise, they do not communicate or signal, for example, a past tense for an action.

Verb phrases are constructed by adding auxiliary verbs to the base verb form to provide subtleties that would make any physicist chuckle since the notion of a past or a future rests largely in our imaginations and in languages (although physics does assume the existence of the past and the future in such areas as the physics of trajectory).

Additional verb forms substantially refine the use of the present, past, or future. The common auxiliary verbs are forms of the verbs *to be* and *to have*. They are the most heavily used of the auxiliaries because they are part of the construction of many verb tenses.

am examining	
is examining	**have** studied
are examining	**has** studied
was examining	**had** studied
were examining	**have been** studying
will be examining	**has been** studying
shall be examining	**had been** studying
could be examining	

Other auxiliaries indicate the *possibility* that something will happen, and others indicate that the verb activity is a necessity or an obligation.

do calculate	**may** alter	
did calculate	**might** alter	**should have** observed
does calculate	**would** alter	**would have** observed
		must have observed
		should have been observed
will predict	**should** conclude	**could have been** observed
shall predict*	**must** conclude	**must have been** observed
can predict		

* In American usage, shall *is mostly used for questions that request confirmation. Shall I open the window?*

The language is highly flexible, and these verb phrases are frequently split when they are used.

Has the satellite **been returned** to NASA?

Have biologists **concluded** that Mars landings will not contaminate the red planet?

Tense

Tense expresses a time frame. Each of the six tenses also has a progressive form to express ongoing activity.

	Basic Form	**Progressive Form**
Simple present:	examine(s)	is (are) examining
Present perfect:	has (have) examined	has (have) been examining
Simple past:	examined	was (were) examining
Past perfect:	had examined	had been examining
Simple future:	will (shall) examine	will (shall) be examining
Future perfect:	will (shall) have examined	will (shall) have been examining

Let's look briefly at the tense logic:

- **Present tense:** happening *now*

I **write** daily.	(basic)
I **am writing** daily.	(progressive)

- **Present perfect tense:** happened *then,* or happened *then* and *still* happening

The star **has collapsed.**	(basic)
The plate **has been falling** at the rate of one inch a year.	(progressive)

- **Past tense:** happened *then*

Alfred Wegener **proposed** the theory of continental drift.	(basic)
He **was living** in his native Germany at that time.	(progressive)

- **Past perfect tense:** event ended *before* another event *began*

Before Lavoisier dicredited the idea, Laurent thought he **had discovered** a substance he called *phlogiston*.	(basic)
The French chemist **had been experimenting** with different concepts of combustion before the principles were understood.	(progressive)

- **Future tense:** not *yet* happening

They **will wear** scuba gear for the dive.	(basic)
They **will be determining** the bearings for the landing.	(progressive)

- **Future perfect tense:** event will *end before* another begins

> The plant **will have bloomed** before July.   (basic)

> They **will have been packing** by the time we return.   (progressive)

Mood

Verbs can state, command, or express conditions contrary to fact.

- The **indicative** mood states a fact.

> Edison's Menlo Park laboratory was only twenty-five miles from New York City.

- The **imperative** mood expresses a command.

> Bring the files to the meeting.

- The most practical use of the *subjunctive* mood expresses an improbable condition or one contrary to fact. It is usually formed by using the past tense subjunctive form of *to be*.

> (if) I were   (if) we were
>
> (if) you were   (if) you were
>
> (if) he/she/it were   (if) they were

The subjunctive mood is not used very often, but you can usually identify where it should be used.

> Ø   If I was you, I wouldn't go.

Because this is a construction contrary to fact—I am not you—the subjunctive must be used.

> If I were you, I would not go.

Voice

A verb can have a grammatical subject that is not the logical subject. In other words, a sentence is usually designed with the subject before the verb, and the subject is understood to be the performer of the verb action (the usual logical practice).

> The sand people collected space junk.

The language structure is so consistent at this basic level that any noun will work as a subject.

> The pears collected space junk.

Here there is a *grammatical* subject; it simply is not a *logical* subject. The grammatical subject is a patterned expectation. It fits the position, but it may not fit the logic.

There is a special use for this peculiarity in the language. Since the grammatical subject (which can be viewed as a *location* in a sentence) can be separated from the logical subject (the *real* performer of the verb), I can create a sentence such as this one:

Space junk was collected by the **sand people.**
    ↑   ↑
grammatical subject   *logical* subject

Here I have reversed the sentence and moved the grammatical object to the subject position before the action. Although commonplace, people do not use this construction with nearly the frequency of the usual situation in which the subject precedes and acts out the verb.

If the grammatical subject and the logical subject coincide, we speak of this as the *active voice*. The subject is in the frontal position. The subject does the job identified by the verb.

The **team** studied the infrared photographs.

If I want to reverse the sentence, I move the logical subject to a position following the verb. This construction is referred to as *passive voice*.

The infrared photographs were studied by the **team.**

A unique aspect of the passive voice is that the logical subject can be totally removed from the sentence!

The infrared photographs were studied.

The small group of *to be* auxiliary verbs is used to structure the passive voice logic. Here is a sampling:

- **Present**

    Transmissions **are** controlled by the satellite.

- **Present progressive**

    Progress **is being** demonstrated by the patient.

- **Present perfect**

    It **has been** projected by the UL study.

- **Past**

    Evidence **was** introduced by the team.

- **Past progressive**

  Input **was being** prepared by the staff.

- **Past perfect**

  Safety **had been** ensured by extensive experimentation.

- **Future**

  Calculations **will be** formulated by the computer.

- **Future perfect**

  Conditions **will have been** improved by the volunteers.

Notice the important function of the word *by* in every example. Turn to the exercise on page 30 and examine the third paragraph. See if you can identify five passive constructions in the paragraph.

To remove passive constructions, you could use the search mode of your word processing software and look for the instances of this preposition. It will signal many of the passives you may have constructed *if* the *logical subjects* remain in the sentences. However, the current grammar check features on your computer are efficient at locating passives and will even suggest alternative phrasing. Though they are sometimes clear and appropriate, the use of passive constructions is often discouraged (see also pp. 98–99 and pp. 178–179) for general stylistic reasons.

Pronouns

Pronouns are a small group of words that occur with great frequency. They function as noun substitutes in endless practical applications. Because pronouns are replacement parts, their relationships to nouns can be troublesome. The personal pronouns are usually monosyllables (for example, the one-syllable words *I, we, he, she, they, you, it*), which suggests their importance in our rapid-fire everyday speech patterns. These pronouns are precise words. There are five other classes of pronouns, however, that can be frustrating.

Pronouns cause considerable confusion, partly as a result of the substitution process and partly because only some of the personal pronouns are exact in meaning: *I, we,* and *she* for example. Many of the pronouns are ambiguous because we do not know to whom or what they specifically refer: *this, these, it, who, whose, which, that, anyone, everyone, nobody, somebody.* Sometimes they do not refer to anyone or anything specifically. Whom or what any one of these words is intended to represent can be difficult to detect. Nevertheless, all of them mean—or replace—a noun that you have to be able to locate and identify somewhere in the sentence or the paragraph or possibly somewhere else on the page! The sentence

  Hand me **that**

will be questioned:

Which?

Demonstrative Pronouns

this   that   these   those

This is the best amplifier.

I could not find the galvanized nails, but **these** will do.

These four pronouns are perhaps the noun substitutes that are the most common and sometimes confusing in our daily conversations, and they invite the questions you know so well: Which amplifier? Which other nails? Which one? That one? These? Those?

Relative Pronouns

A relative pronoun is the first word of an adjective clause, one that functions to modify a noun. A number of the pronouns identified previously can function as relative pronouns.

who, whose, that, what, whoever, whichever, which

They function as relative pronouns when they specifically serve to initiate an adjective (or relative) clause.

The space probe, **which NASA had pronounced a failure,** began to send back photographs two weeks later.

The surveyors brought the laser **that they needed.**

Indefinite Pronouns

The indefinite pronouns are the largest group of pronouns, and they include a number of words that function as adjectives also.

all	**everybody**	**no one**
another	**everyone**	**nothing**
any	**everything**	**one**
anybody	**few**	**other**
anyone	**many**	**several**
anything	**most**	**some**
both	**much**	**somebody**
each	**neither**	**such**
each one	**nobody**	
either	**none**	

Everyone went to the demonstration.

Nobody stayed in the lab that night.

Someone had taken **something**.

I had **some**.

I had **several**.

There were a **few**.

Interrogative Pronouns (for questions)

who   whose   what   which

What was the objective?

Who was there?

Emphatic Pronouns

Authors often add the endings *-self* or *-selves* to pronouns to produce emphasis. These are also called *intensive* or *reflexive pronouns,* and the former term seems to appropriately explain their use.

Fermi **himself** was placed in charge of the Manhattan Project.

The scientists **themselves** were shocked at the power of the first hydrogen bomb, which was 500 times more powerful than the World War II blasts.

Reflexive pronoun forms also can be used to refer to someone or something in a sentence to show that the person or thing is acting on or for itself.

The car was so sophisticated it could almost drive **itself**.

In the asylum where he worked, few patients could feed **themselves**.

In fact, care had to be taken to make sure one patient would not choke **himself**.

Adjectives

An adjective describes a noun (or a pronoun) or another adjective. The adjective is said to "modify," meaning that the word defines a condition for a noun or adjective. It defines the noun or adjective. When nonnative speakers learn English, they are taught that English speakers put adjectives before the nouns. That is not always true. Adjectives can be put after nouns, as in French. They can even be put elsewhere in the sentence.

The aromatic green limes sat in the sunlight.

The limes, aromatic and green, sat in the sunlight.

The limes sat aromatic and green in the sunlight.

Predictably, an adjective will answer one of several questions:

- What kind?

    An **effective** leader is a good listener.

    Sanitary landfills still dispose of 80 percent of our solid waste.

    Medical incinerators emit high levels of toxins.

- How many?

    There were **eight** dogs in the pack.

    Ten million species may inhabit the earth.

    Eighty-seven countries now participate in the endangered species treaties.

- Which one?

    That gun went off.

    The **soil's** mineral ions must be absorbed by the plants.

    Our lakes are often damaged by acid rain.

An adjective is defined by its function and not necessarily by its usual capacity as a part of speech. For example, a number of pronouns will explain "which one." Obviously, possessives do the job by stating ownership. The demonstrative pronouns (*this, that, these, those*) will explain "which one," as well as the articles *a, an,* and *the.*

> **This** incident is **an** example of **the** risk involved in **their** space flight.

Proper nouns clearly identify "which one."

> The **Porter-Cable** saw worked well.

> I use **PageMaker** software for my projects.

In other words, whereas a noun is usually a noun, several parts of speech can function as adjectives. Adjectives often also describe by degree, which is evident in our daily use of comparison adjectives. The comparative form of an adjective compares two things, and a superlative form compares three or more.

Base	*Comparative*	*Superlative*
good	**better**	**best**
tall	**taller**	**tallest**
heavy	**heavier**	**heaviest**

Articles

The articles consist of the three words *a, an,* and *the.* They are also known as limiting adjectives, although they are probably logically redundant in most usage.

dog

a dog

the dog

These monosyllables are used with intensity regardless of the curious lack of logical necessity for the devices. In fact, many languages do not use articles. English speakers use the articles to make one critical distinction. *A* towel (an indefinite towel) is not quite the same as *the* towel (a definite towel). We use the articles to distinguish the general from the specific.

Adverbs

An adverb is another part of speech that modifies; that is, the adverb details the specific conditions of a verb, another adverb, or an adjective. The adverb is well named, since it is usually a word *added to* a *verb.*

Waksman **successfully** located the first bactericidal compound in the streptomycin group.

The data proved **very** significantly that there was a likelihood of an earthquake.

The doctors cured **considerably** fewer cases after the bacteria adapted to the drugs.

Because adjectives and adverbs are the two primary groups of "descriptors," or groups of words describing words, they often are variations of the same words, possibly originating as adjectives since the usual *-ly* endings of many adverbs were clearly added after the original words were created. It is generally true that the *-ly* ending often identifies adverbs.

The children were very **proper.**   (adjective)

The children behaved **properly.**   (adverb)

There was a **sudden** change.   (adjective)

They changed **suddenly.**   (adverb)

There are many adverbs that do not end in *-ly.*

I work **hard.**

He will go **far.**

It was falling **fast.**

They went **straight.**

Like adjectives, adverbs answer questions because they describe.

Adverbs of Time

- When is the action?

    I have to go **now** to record the test.

    A solution will **eventually** follow.

    Call the doctor **immediately.**

    Trauma incidents refers to accidents that occur **suddenly.**

Adverbs of Place

- Where is the action?

    Push the table **backward.**

    The rocket fired **downward** by mistake.

    The shoots of horizontal stems will grow **upward** because of apical dominance.

Adverbs of Method

- How was the action done?

    The results of the investigation were **bitterly** challenged.

    They prepared the blood samples **cautiously.**

    They did **not** perform the proper road tests.

- What is the intensity of the action?

    They **thoroughly** examined the data.

    The astronomer Karl Schwarzschild **brilliantly** proposed the black hole theory.

    The crater was **far less** interesting than we expected.

    The exercise was not supposed to be performed **vigorously.**

In the preceding samples some of the adverbs could be placed in several categories: *suddenly* (time and method), *thoroughly, brilliantly, vigorously* (method and intensity).

Prepositions

The prepositions are an important and specific vocabulary for orienting nouns in terms of time and space. Here is a list of the prepositions that we use to describe or position sentence logic in terms of time or location:

Common Prepositions		
about	beyond	past
above	by	pending
across	concerning	regarding
after	down	since
against	during	through
along	except for	throughout
amid	from	to
among	in	toward(s)
around	inside	under
at	into	underneath
before	near	until
behind	off	unto
below	on	up
beneath	onto	upon
beside	out	with
between	outside	within
	over	without

No other part of speech serves quite the same function for providing bearings. The prepositions "position" our mental image or the visual composition that a sentence constructs. (One exception is the preposition *of:* He is one *of* the best).

> He immediately looked **through** the telescope instead of using the viewfinder.
>
> There, **inside** the cell, was the cause of the explosive and unexplained growth we call a malignancy.

Conjunctions

Words, phrases, clauses, and sentences can be joined with a small group of words we call conjunctions. They are of three types.

Coordinating Conjunctions

and   **or**   **but**   **for**   **so**   **yet**   **nor**

These are daily fare, and the first three are used with considerable frequency to join words, phrases, and sentences.

Radio telescopes **and** computers are at the heart of the Very Long Baseline Array system.

Teller **and** other scientists built the hydrogen bomb, **but** Oppenheimer objected to the dangers of such a weapon.

The moon rock samples posed a contamination risk, **and** NASA was not effective in establishing isolation procedures for the samples.

The original London Bridge was composed of 130,000 tons of stone, **but** that was the weight of only the exterior framework.

Conjunctive Adverbs

These are the larger connectors that join sentences (independent clauses).

however	**nonetheless**	**hence**	**finally**
therefore	**nevertheless**	**consequently**	**otherwise**
accordingly	**indeed**	**furthermore**	**besides**

The first birth control pill was marketed in 1960, and early versions contained 100 to 150 micrograms of estrogen; **however,** subsequent research indicated that 50 micrograms was the desirable dosage!

Correlative Conjunctions

These connectors work in pairs.

both	. . .	**and**	. . .
either	. . .	**or**	. . .
neither	. . .	**nor**	. . .
whether	. . .	**or**	. . .

During the 1970s energy crisis, politicians ignored **both** solar radiation **and** geothermal power.

Subordinating Conjunctions

A fairly large group of conjunctions can be used at the beginning of a dependent clause to create the subordinate structure.

after	if	than
although, though	in order that	that
as	in that	unless
as if	inasmuch as	until
as long as	now that	when
as much as	once	where
because	provided that	whereas
before	since	wherever
even though	so long as	while
how	so that	whether

The independent clause is a sentence that can stand alone. The dependent clause is relational because of the subordinating conjunction that precedes it. It cannot stand alone. The following samples are independent:

> I went to the bank; I had to make a deposit.

> Building fires are not hot enough to melt steel structural members; however, fire can weaken the metals sufficiently to cause collapse.

Introducing a subordinating conjunction changes the construction and makes the second clause "dependent" on the first.

> Eiffel built his famous tower on beds of concrete seven feet thick **because** the wrought iron structure weighs 7300 tons.

> An asteroid may have caused the disappearance of the dinosaurs 65 million years ago **if** there was enough dust created by the impact to temporarily block sunlight around the globe.

Interjections

Ouch!   **Wow!**   **Yes!**   **No!**   **Yuk!**

Though commonplace in our speech patterns, these informal elements are omitted from formal or professional and technical writing.

Verbals

Strictly speaking, verbals are not one of the basic word groups that make up the parts of speech. Because subjects and verbs are the engines of the language, it is quite appropriate to use them in a number of capacities. The most interesting group involves word versions of "freeze-frames" (produced by the stop-action feature on a video system). Like a hummingbird photographed in flight, verbs can be changed to nouns forms called verbals to make them stand still.

Verbals are words derived from verbs, but they do not function as verbs. Instead, they act as nouns or adjectives or adverbs. The concept is there, but the action is not.

I run.	He hunts.
I like **to run.**	He enjoys **hunting.**

There are three types of verbals.

Participles

The present participles and the past participles of verb forms can function as adjectives.

He asked for **revised** calculations for the second meeting.

The **exploding** fireworks were a thrill.

Increasing ultraviolet radiation resulted from the ozone hole.

The **fallen** leaves were beautiful because of the carotenoid pigments.

The **printed** copy came from his computer.

She calculated an **appraised** value.

The **broken** record could not be restored.

Gerunds

The other two types of verbals are the freeze-frames made by moving the verb into a noun capacity to force a shift from the dynamic state to the static state.

The gerund is the present participle (the verb form with the *ing* ending) used as a noun.

Weight lifting is growing in popularity.

Exercising is rewarding if the activity is aerobic.

Highway **engineering** looks a little dull until you reach the Rockies or Key West.

Infinitives

The infinitive form of a verb often functions as a noun or adjective or adverb.

To win the superconductor race we must spend billions; **to lose** the race will cost us billions.

Five commands are the keys **to use.**

The extra speakers helped **to balance** the sound, but quadraphonics seemed artificial for solo instrumentation.

Exercises Chapter 4, Section C

1. Underline the nouns and the adjectives in the following paragraph.

After the discovery of the photoelectric effect in the 1870s, curious engineers and daring inventors began to develop photoelectric cells to transmit images. The scientific concept was correct, but the complicated approach was unrealistic because each little dot of an image was a cell with a wire that had to go to the distant transmitter.

2. Underline the verbs and the adverbs in the next paragraph.

Experimenters eventually developed the amplifying triode tube, and this device made the original system possible, but the hundreds of wires were extremely inconvenient. Researchers then proposed a signal array technique that we now know as scanning. The scanning concept worked well and led to a variety of devices in the following decades. The first "television" broadcasts were made in New York in 1928, but the signals were badly transmitted by today's standards.

3. Underline the pronouns and prepositions in the following paragraph.

A Russian-American physicist, Vladimir Zworykin, developed the first reliable scanning system for the budding television industry. After extensive experimentation, he built an effective scanning device. He used a cathode ray tube (CRT) in the camera and another one in the receiver. He had patented a television tube (he called it a "kinescope") in 1924, but critics were probably skeptical about the device. He proved them wrong in 1938, *when, after adding improvements, he produced a working model of the entire technology. During World War II and throughout the 1940s, military research on CRTs soared because the Allies realized the importance of radar and sonar in warfare. Following the war, engineers adapted the war technology to television, and they soon brought down production costs. Consumers demanded television sets, and they were excited about the new technology because television was suddenly affordable.*

---

4. *Locate the* coordinating conjunctions, subordinating conjunctions, *and the* conjunctive adverbs *in the final paragraph.*

Color television slowly gained popularity, but the cost of a set was prohibitive. Like the HDTV technology of today, early color televisions were pricey; however, mass production costs and consumer drives soon made the systems affordable. The color technology was partly based on the original black-and-white processes, although it was complicated by the three-color scanning method. The three signals are beamed together, and the receiver separates the component parts. Unlike the black-and-white screen, the color screen is coated with color-sensitive phosphors; therefore, they respond in color. As a result, the viewers can watch their favorite shows in color. Television is perhaps the invention of the twentieth century. Indeed, the word was coined in the year 1900, and HDTV was in the showrooms before the year 2000.

Spelling Strategies

There are two approaches to spelling; the first is the electronic highway and the second is the traditional set of rules that has always governed spelling practices. Let's look at the modern realities first.

- Always use your spell checker. Be sure you use American spellings if the software provides you with options.

- Key in the vocabulary that is critical to your specialty. The spell checker component of a word processor holds a generous but generic list of words. Do not expect scientific terms to be in a spell checker simply because they have been around awhile. *Prions* were discovered in the 1920s, and *viroids* were discovered in the early 1970s, but neither of these microbiological terms is in my spell checker or my dictionary. Technical terminology must be imported. If you consistently add new words, you will have a handle on spellings that meet your specific needs.

5

- Keep a desktop spell checker on hand. They can cost less than $30, and they include search features for really bad spellers.

- Consider buying a dedicated program for your discipline if you can locate one. These programs may have a guide to terminology, and they may have a spell check application that is dedicated to your discipline.

- If, however, you dump your laptop in a stream while you are on a fishing trip, there are basic rules for all but a handful of exceptions.

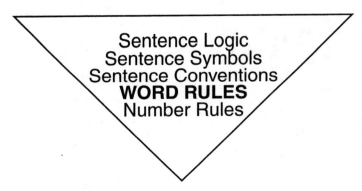

Sentence Logic
Sentence Symbols
Sentence Conventions
WORD RULES
Number Rules

English is the fastest-growing language in the world, and most of that growth is the result of scientific and technological advances. There probably is not much likelihood that either your spell checker or the traditional rules will help if you have to dash off terminology on the order of *optoisolators, praseodymium dopant, errmsg parameters,* or *schlieren system.* On the growth end, the language can be very complicated when there are subjects like these from NASA technical reports:

Integrable Photonic Devices by Selective-Area Epitaxy

Dynamically Reconfigurable Optical Morphological Processors

The misspellings I find in technical reports composed by students usually involve about five types of errors. The modern terminology of scientific activity is one of them. At the opposite end are the historical spellings of a very old language and the borrowings that an old language gathers along its course of development. Over a period of a thousand years, quirky patterns become part of a language. In English, old words such as *knife* and *know* are typical oddities because of the presence of the silent *k,* and there are hundreds of others. The problem is that they frequently are part of our daily vocabulary a thousand years later.

Borrowings are also pesky. The French government has been hard put to stop the influx of English computer terminology, which has been adopted by *les hackers* and *les surfers* of France. Some of the most confusing words Americans try to spell in English, however, are terms that were borrowed from the French long ago. Military terms—*sergeant, battalion, lieutenant, liaison, rendezvous, camouflage*—are common. Cooking terms are ever popular also: *bouillon, hors d'oeuvre, sauté, gourmet, mayonnaise, hollandaise, beef bour-guignon crème fraiche.* These are fairly common-place terms. English borrowed hundreds of other French words from *bureau* and *clientele* to *repertoire* and *résumé.* The problem is that old language and borrowed language often defy the spelling rules. As students often comment, "I cannot find *naive* in the dictionary because I cannot spell it!"

The fourth category is perhaps the most common of the groups of errors that I encounter: the vowel errors. Years ago I would often hear the old line, "If I can't spell it I can't find it in the dictionary to check the spelling." I realized that the answer to this catch-22 chal-lenge is a little patience. When you look for a misspelled word and cannot find it, try the vowel sounds. Americans twist a number of their vowels into a generic "uh" sound and, for example, "the" becomes "thuh." With a little patience, you will usually find the spelling if you try *a, e, i, o u,* even though *naive, pneumonia,* and others will challenge your patience.

The final category of spelling errors I often see involves the various endings writers try to attach to words. The *ly* endings are confusing to some writers who accidentally spell words incorrectly at the syllable that joins the *ly* ending. My rule: simply be ready for trou-ble if you have an *ly* ending, and look it up or spell-check it.

approximately	subtly	incidentally	disastrously
immediately	truly	successfully	exceedingly

Plurals

- Add *s* or *es* to most nouns to construct plurals. The single *s* is far more common.

tunnels	drains	turbines

- Add *es* to words ending in *s, sh, j, x, z.*

rhinoceroses	squashes	taxes
processes	wishes	faxes

- Drop the *y* and add *ies* to form the plural of words ending in *y* if a consonant pre-cedes the *y.*

lady	→	ladies	day	→	days
malady	→	maladies	monkey	→	monkeys

- If a word ends in *o,* add an *s* unless a consonant precedes the *o.* Then end the word in *es.*

video	→	videos	tomato	→	tomatoes
rodeo	→	rodeos	zero	→	zeroes

There are always exceptions to the *s* plurals, of course. Two examples are

woman → women mouse → mice

Contractions

An apostrophe replaces missing letters in a contraction. This is not exactly a spelling error, but it is often misplaced within a word (see also p. 79).

wouldn't couldn't can't won't

Since many writers need to be more formal in their writing, one immediate way to tighten up a text is to remove contractions, which will also remove less desirable expressions such as *what's* and *he's*.

Use contractions if you construct a question, because the uncontracted forms can sound stilted. That is, the question

Can you not download the material?

should be

Can't you download the material?

Similarly,

Do you not like the new TI calculators?

should be

Don't you like the new TI calculators?

Divisions

- If you use a hyphen to divide a word at the end of a typed line, be sure the word is divided between syllables.
- Avoid end-of-the-line hyphens. Justify the right margin to reduce the occurrence of hyphens, or, if typing, allow the right edges of the typed lines to vary in length rather than hyphenate.

Compound Adjectives

- Hyphenate word clusters that function as single adjectives when they occur before a noun, but omit the hyphens if the adjectives follow the noun.

She was a **middle-aged** woman.

The woman was **middle aged.**

It was an **off-the-wall** comment.

The comment was **off the wall.**

- When authors fail to see compound adjectives, they can create confusion.

  Ø   About 25,000 people a year die in driving while drinking accidents.

This sentence is easy to understand if it is punctuated correctly.

About 25,000 people a year die in **driving-while-drinking** accidents.

Finding your compound adjectives can be a drawn-out, time-consuming process.

Prefixes

- Do not use hyphens when forming compounds with the following prefixes. These prefixes are attached to an enormous number of technical words: *megabyte, transducer, semiconductor, infrared, interface, protoboard, minicomputer.* There are exceptions, however, such as *un-ionized* (not ionized), *re-create* (create again), *recover* (cover again).

ante	intra	non	socio
anti	macro	neo	sub
bio	mal	over	super
co	mega	post	supra
contra	meta	pre	trans
counter	micro	pro	ultra
extra	mid	proto	un
hypo	mini	pseudo	under
infra	multi	re	
inter	mis	semi	

There are many *quantitative* prefixes that are used in the development of scientific vocabulary: *uni, bi, tri, multi, milli, omni, non.* From *anti*matter to *micro*chip to *ultra*sonic, prefixes have become part of thousands of new scientific terms in the last hundred years.

- Use a hyphen after the following smaller group of prefixes, many of which are English words: *all-, ex-, half-, quarter-, quasi-, self-.*

  all-encompassing   quarter-round molding

  ex-director   quasi-typical

  half-time instructor   self-directed

Suffixes

Words ending in Silent e

- Drop the silent *e* when adding suffixes that begin with a vowel, for example, *-ing, -able, -age, -ed, -ence, -ent, -ist.*

staple	→	stapling
shade	→	shading
engage	→	engaged
adhere	→	adherent
feminine	→	feminist

- If there is no silent *e*, simply add the ending.

leverage	sanded	transportable
dependent	tourist	blending

- If the suffix begins with a consonant, add the ending to words that end with a silent *e* without dropping the *e*.

polite	→	politely
puzzle	→	puzzlement
comfortable	→	comfortableness

Words Ending in Y

- Adding a suffix to words that end in *y* sometimes creates confusion for writers. The rule is that if *y* is preceded by a consonant, change the *y* to *i* before adding the suffix. If the *y* is preceded by a vowel, add the suffix to the existing word.

totality	→	totalities	tragedy	→	tragedies
reality	→	realities	pry	→	pries
discovery	→	discoveries	hazy	→	hazier
liberty	→	liberties	frisky	→	friskiest
play	→	played	employ	→	employable
stay	→	stayed	betray	→	betrayed

As a curiosity, notice that the *-y* ending itself is often a suffix. That is how *victor* became *victory,* for example.

- Do not change the *y* if the suffix is *ing*.

spy	→	spying
supply	→	supplying

Double Consonants

- If a one-syllable word ends in a consonant, double the consonant before adding a suffix that begins with a vowel.

    Do not pet the animals; the sign says, "No **petting.**"

    They were **stunned** by the **hidden** alarm.

- Similarly, if a consonant is preceded by a stressed syllable in a word with more than one syllable, double the consonant.

    Any **occurrence** of this type of error is **forbidden.**

I before E

Everyone knows the familiar rule that *i* goes before *e* except after *c*. Older rule books state the *i*-before-*e* rule as "*I* before *e* except after *c* and when rhyming with *a* as in *neighbor* and *weigh*."

conceive	perceive	receive	deceive

Once in awhile, you will see an exception that may look a little *weird* (for example). The word *science,* and others like it, are also notable because the *i* and the *e* are split into two sounds in two separate syllables: *sci-ence.*

If you use exceptions such as *either, neither, height,* or *coefficient* frequently in your writing vocabulary, memorize them with a little witty word play. That way, remembering how to spell *believe* will be a *piece* of *pie.*

Rhyme Spellings and Look-Alikes

By far the greatest number of spelling errors involve rhymed words (homonyms) and words that look or sound similar to one another. *To, too,* and *two* are the most obvious examples. Other are listed below. The spell checker will not identify any of these words as your spelling zingers, and yet they are the most persistent errors. What to do?

I suggest that you identify those rhymed spellings that are your frequent errors. After you spell-check a document, enter a search for, let's say, *whether* and *weather,* and check the text. (The nonexistent *wheather* will show up in the spell check.) You will be able to identify most of these blunders as wrong if you apply a little scrutiny because the word itself will be incorrect. The spelling is a secondary error. Unfortunately, the *to-too-two* group will take a while to proofread because the words appear very frequently in writing.

Here are the most common of the rhyme errors and look-alikes organized somewhat by the frequency with which they occur in projects that you will edit.

to, too, two	advice, advise
their, there, they're	maybe, may be
weather, whether	than, then
affect, effect	vary, very
capital, capitol	all ready, already
who's, whose	all together, altogether
principal, principle	coarse, course
accept, except	of, off
cite, site, sight	everyone, every one
farther, further	awhile, a while
its, it's	anyone, any one
lead, led	your, you're
loose, lose	sometime, some time, sometimes

The following rhyme pairs and look-alikes are less frequently used, but it is important to know their meanings. Misuse of the rhyme pairs indicates a vocabulary problem rather than a spelling problem.

adapt, adopt	eminent, imminent
allude, elude	forward, foreword
allusion, illusion	later, latter
always, all ways	past, passed
beside, besides	precede, proceed
born, borne	purpose, propose
break, brake	quiet, quite, quit
carat, carrot	roll, role
conscience, conscious	sense, since
continual, continuous	though, thought
dyeing, dying	through, thorough
elicit, illicit	

At times, the rhyme will create words that do not exist: *wreckless* or *fowled*. Then you have a spelling error, rather than an error in word choice.

-ed Endings

As noted earlier, the past tense and past participle of most verbs take the *-ed* or *-d* ending. In nonstandard spoken English, the "d" sound is often dropped, and even well-educated speakers sometimes find it difficult to add the *d* to a few popular expressions:

used to	*not*	"use to"
supposed to	*not*	"suppose to"
prejudiced	*not*	"he is prejudice"
renowned	*not*	"renown leader"

These mispronunciations can become misspellings. You might try to remember that *use* will need a *-d* ending if the word is followed by *to*.

A Few Term Clarifications

The rhyme pairs can be divided into three sets. The frequent spelling or vocabulary errors (such as *to, too,* and *two*) constitute the group I call "zingers." Likely bobbles of spelling or usage errors (like *past, passed*) are the second group that authors often have to look up in the dictionary. A third group involves the few rhymed pairs or look-alikes that may play a role in technical documentation. This third sort of housekeeping is best kept on the search list of your technical vocabulary. The terms that are often misused in technical work include the following:

- **activate, actuate**

  Both terms are acceptable in the sense of *to start,* although *start* is usually more appropriate than either term.

  > The power-on button **actuated** the apparatus.

- **affect, effect**

  Affect is most commonly used as a verb meaning to influence.

  > How will the change **affect** you?

  Effect is both a noun and a verb, but is most commonly used as a noun.

  > What is the **effect** of adding the solvent?

- **analysis, analyze**

  Analysis is the noun.

  > There was only one **analysis.**

  Analyze is the verb.

  > I can **analyze** the smear.

- **cite, site, sight**

  Cite is the verb, meaning *to acknowledge* or *to quote*. *Sight* is vision. *Site* is the location. The word of choice for most engineering applications is *site*.

  > The conduit was moved to the construction **site.**

  > This was the **site** of the explosion.

- **criterion, criteria**

  Criterion is singular.

  > The judgment was based on one **criterion.**

  Criteria is the plural.

  > The **criteria** were not established.

- **datum, data**

  Datum is singular.

  > The simplest **datum** would help the investigation.

  Data is plural, and is usually the word you will want.

  > The data **were** easy to compile.

- **formula, formulas, formulae**

  Formula is the singular.

  > One **formula** was correct.

  Formulas is the plural.

  > There were two **formulas.**

  Formulae is the Latin form. Do not use it. Words such as *antennae* have more or less vanished even though the -ae ending does show up occasionally in a plural such as *vertebrae*.

- **medium, mediums, media**

  Medium is singular.

  > Air is the **medium** for sound energy.

Mediums is the more practical and the more widely used plural.

> Thanks to modern chemistry, the three new **mediums** gave artists alternative avenues for artistic expression.

Media is the plural form used in reference to specific industries such as telecommunications.

> CD-ROM technology will create new electronic **media,** as well as new avenues for the entertainment **media.**

- **per cent, percent, percentage**

Use the word *percent* for percentages that are stated in words.

> fifty **percent**

The two-word form *per cent* is an older version. Use the one-word form.

Use *percentage* when no numbers are involved.

> The **percentage** of error was minimal.

If the percentage is expressed in numbers, use the symbol %.

> 20%

- **phenomenon, phenomena**

Phenomenon is the singular.

> The UFO incident is an unexplained **phenomenon.**

Phenomena is plural.

> Atmospheric **phenomena** are used to predict the weather.

- **stratum, strata**

Stratum is singular.

> There is only one **stratum** of enamel on the tooth.

Strata is plural.

> The **strata** were evident because of the erosion.

Other Spelling Errors

- **existence**

  Observe that the word contains only *e*'s in addition to the *i*.

- **nuclear**

  This word is frequently mispronounced as "nucular" but fortunately this spelling is so odd that it is not likely to be transferred to paper as a spelling error.

- **separate**

  The word is spelled with a pair of *a*'s between a pair of *e*'s.

  > An old tip: There is "a rat" in *separate*.

- **solder**

  In this case the correct pronunciation is quite unlike the spelling, but few people would recognize the word without the *l*. Similarly, *caulking* is frequently pronounced *without* the *l*. Less conspicious are common words such as *talk* and *walk*.

- **species**

  Both the singular and plural forms are spelled *species,* with a final *s*.

Exercises Chapter 5, Section A

Correct the errors in the following sentences.

1. You could hear the echos in the valley.

2. The research is viable if she's using the right variables.

3. Follow the step by step instructions.

4. Semi-conductor technology began in the 1950s.

5. Meta-engines are a recent Internet trend.

6. Infra-structures have been greatly improved because of progress in metallurgy.

7. He was usualy confused from the begining.

_____

8. Early scientists beleived the planets went around the earth and that their were 6.

_____

9. The meteor fell near this cite.

_____

10. The piston cracked because the rings were lose.

_____

11. The Doppler affect was a principle demonstration that the universe was expanding.

_____

12. A quiet car is dangerous at high speeds because it creates the allusion that a driver will be able to break quickly if the occassion arises.

Capitalization Standards

At one time, capitalization was largely a matter of an author's style, and frequent and non-standard uses of caps were common. In today's academic and technological settings, rules are more the norm, and the rules limit what some authorities call "excess capitalization."

Nowadays, capitalization is largely used to distinguish a specific thing (or person or place) from a generic variation or grouping of the thing or things.

> The members of the Republican Party believe in the democratic values of the American culture.

Capitalization is a social practice more than a grammatical issue. Capitalization is a convention that has evolved out of social traditions. Specific entities such as town, country, or state names, or park and monument names emerge from the historical context of the society that produced them. Similarly, cultural values such as status have often been rendered with capitalization, of family names and titles, for example. The tradition lives on when a veterinarian is identified as *Paul Henderson, D.V.M.*

Personal and Group Names

- Capitalize *personal* names and initials.

    D. R. Wallace

    Susan B. Anthony

- Capitalize *titles* used with names.

    Professor John Truman

    Dr. Richard Hare

    Governor Cooper

- Capitalize the names of degrees if cited after a personal name. They may be abbreviated or occasionally written in full. Do *not* capitalize general references to degrees.

    Barbara Dill, **M.D.**

    Dr. Carl Anderson, **D.D.S.**

    He was a **doctor** of veterinary medicine.

- Capitalize titles that replace the name in a direct address of a person.

    Park the plane, **Lieutenant.**

- Do *not* capitalize titles that are used as common nouns.

  The **board of directors** meets this evening.

  We did not meet the **attaché** at the consulate.

  The **biologist** was the last man on board.

- Capitalize kinship titles that replace the name in a direct address.

  Look, **Dad,** I can't afford the tuition.

- Do *not* capitalize kinships that are indirect references.

  His **father** wanted him to go to college.

- Capitalize the names of racial, national, tribal, ethnic, regional, or similar groups.

  Hispanic   African American   Chicano

- Do *not* capitalize references to *social classes,* however.

  working class   blue-collar employees

Place Names

- Capitalize recognized geographic names of cities, towns, rivers, mountains, and other distinguishing features commonly identified on maps or in literature.

  the Amazon River   South America   Cordoba, Argentina

- Capitalize these same identifications when they are used as adjectives in the descriptions of other place elements.

  the Amazon basin

  South American political difficulties

- Do *not* capitalize, meridians and parallels or the word *equator.*

  The forty-eighth parallel borders Canada on our northern border.

  I fly over the equator twice a year.

- Capitalize the names of important architectural structures and urban features. The importance may be international or quite local. Include buildings, bridges, monuments, squares, parks, streets, roads, and so on.

  Falling Waters   Telegraph Hill

  Smith Tower Building   Pine Street

Green Lake	Pont Neuf Bridge
McKeesport Park	West Seattle Viaduct

Time Frames

- Capitalize the days of the week, the months of the year, but *not* the seasons.

Monday	October	fall	spring

Organization Names

- Capitalize major agencies, departments, bureaus, offices, and other specific service areas of government, and capitalize professional organizations and businesses.

    Phoenix Housing Authority

    Pasteur Institute

    Department of Infectious Diseases

- Capitalize the popular abbreviations and acronyms of such agencies and of companies.

OSHA	IRS	IEEE	IBM
FDA	NASA	UL	GE

- Similarly, capitalize the names of corporations, associations, unions, and other such groups, as well as their officers and offices.

    International Brotherhood of Electrical Workers

    U. S. West, Foreign Services Division

    American Rose Society

    National Science Foundation

Cultural Names

- In general, capitalize historical ages and other commonly recognized historical periods (see also p. 152).

    la Belle Epoque

    the Paleozoic Age

    the Roaring Twenties

- There is one exception: do not capitalize the period if it is designated by a number.

the nineteenth century

the thirties

- In general, capitalize well-known events, including wars and battles.

    Labor Day

    Battle of the Marne

    World War II

    Cinco de Mayo

- Capitalize awards.

    Fulbright Award   Academy Award

    Nobel Prize   Cannes Film Festival Award

- Capitalize the names of religious groups, their members, the names of their god or gods, their sacred texts, parts of those texts, and sacred events.

    Methodism

    Roman Catholic

    God

    Tao de Ching

    Diaspora

- Capitalize adjectives that derive from proper names. These are usually the names of well-known authorities or acknowledged personalities of some fame.

    a Freudian text

    Jungian analysis

    Ptolemaic astronomy

Titles of Written Works

- Capitalize key words in the titles of books, plays, magazines, articles, pamphlets, proposals, reports, and similar documents. Specifically, capitalize the first word, nouns, verbs, and other descriptive words. Do *not* capitalize articles (*a, an, the*), common conjunctions (*and, but, or, for, nor*), prepositions (*over, beside, under, in,* and others), or the *to* in an infinitive (such as *to explore, to research*), unless it is the first word of the title.

*Infrared Mapping **and** Satellite Robotics*

*Stress Crazing **of** Polymer Sheets **under** Pressure*

*Affordable Nebulizers **to** Clean Particle Counters*

- In general, render titles as they appear in the written materials you cite. If they are in full caps, however, then you must type the title in uppercase and lowercase.

- You may use all caps for your own title on your cover page, but you must type it in uppercase and lowercase if, for example, you cite your document in a cover letter.

Cover page rendering:

THE DESIGN OF THE AXIAL COMPRESSOR BLADE FOR AIRCRAFT ENGINES

Cover letter rendering:

I have enclosed the report *The Design of the Axial Compressor Blade for Aircraft Engines.* This project was completed last year at our facility here in Houston.

- Do *not* use full capitalization for titles in a bibliography.

- Do not capitalize the words for sections of documents, as in *preface, introduction, appendix, index,* and *chapter.* Capitalize document sections only if you are cross-referencing by title to a *specific* section, for example, *see Chapter 3.* The numbers are always in Arabic unless the original document was unconventional and used Roman numerals or some other format.

Other Media

Capitalize titles of motion pictures, radio and television programs, song titles, classical compositions, paintings, sculpture, CDs, software, and related media.

Scientific Terminology

Binomial ("two-name") nomenclature is the familiar, but possibly confusing, method of naming all living things, from bacteria to elephants, using terms generally based on Latin and Greek.

Homo sapiens

Saccharomyces cerevisae

Yersinia pestis

- The first name designates the *genus,* and its first letter is capitalized. The second name is the *specific epithet,* and it is not capitalized. Together the genus and specific epithet identify the *species* of the organism. Both names are italicized.

| genus | → | Rosa | genus | → | Mus |
| species | → | Rosa moschata | species | → | Mus musculus |

In general, the genus is usually abbreviated after first use.

> *R. officinalis* is the apothecary rose.

- The kingdom, phylum, class, order, and family are capitalized.

Kingdom	→	Plantae	Kingdom	→	Animalia
Phylum	→	Anthophyta	Phylum	→	Chordata
Class	→	Dicotyledones	Class	→	Mammalia
Order	→	Fagales	Order	→	Rodentia
Family	→	Fagaceae	Family	→	Muridae

- Common names of plants and animals are irregular. Some are capitalized, but many are not. Historical practice is your guideline. Use a dictionary for standards. Broader categories are generic, so breeds of animals and varieties of plant groups are usually not capitalized.

> grizzly bear   Emperor penguins
>
> Saint-John's-wort (hypericum)   lizards

- Commonly used English adoptions of scientific names are not capitalized or italicized.

> cambrium   protozoa   feline
>
> quadriceps   deltoids   cerebrum
>
> estrogen   meningitis   tendinitis
>
> enzyme   herbivore   anorexia

- Geological time frames are capitalized, but not the words *era, period, epoch, age,* and the like.

> Mesozoic era   Paleocene epoch
>
> Cambrian period   Proterozoic era

- Unique astronomical bodies are capitalized except for generic astronomical terms that form part of the name. Generic references are not capitalized.

> Phobos and Deimos are the moons of Mars.
>
> Crab Nebula
>
> Donati's comet

Medical Terms

- Most medical terminology is not capitalized or italicized. The terms include diseases, procedures, anatomical parts, and a host of biological and microbiological features in the many specialty areas of medicine.

dorsal	thoracic	endoderm	aorta	cytoplasm
interstitial	pineal	vertebrae	pharynx	prolactin

- Popular acronyms are often capitalized. Acronyms are abbreviations that are popularly adopted as a spoken word, such as laser or ASCII or SCSI (pronounced "scuzzy") drive unit.

CAT scan	TURP procedures
SMAC test	MAR records

- Popular abbreviations are capitalized.

IV	H&P	CBC	CXR
VA	MRI	PCP	SGA

- A distinction is made between a disease and the pathogen that causes it. The disease is not italicized, but the pathogen is because it falls under the rules for binomial nomenclature.

chlamydial infections	*Chlamydia psittaci*
Legionnaire's disease	*Legionella pneumophila*
tuberculosis	*Mycobacterium tuberculosis*

- Brand names of medical products and pharmaceuticals are capitalized, and generic names are not.

  The therapist prescribed aspirin or one of the newer anti-inflammatories such as Aleve.

Math and Chemistry

- Only proper names attached to theorems, principles, laws, and similar major truths that emerge in the sciences are capitalized.

Doppler effect	Gödel's proof	Planck's constant
convergence theory	law of conservation	special theory of relativity

- The names of chemical elements and compounds are lowercased when they are written out; however, the chemical symbols are capitalized.

---

helium	He	sulfuric acid	H_2SO_4
radium	Ra	nitric acid	HNO_3

Subscripts and supercripts are frequently used with chemical symbols. A subscript following a symbol indicates the number of atoms in a molecule.

Ozone (O_3) is quite toxic.

Similarly, the mass number is a superscript to the left of a chemical symbol.

$$^{238}U \qquad\qquad ^{14}C$$

In popular writing the mass numbers follow the written name of the element.

Reactor pellets of uranium dioxide are only 3% **uranium-235.**

In older literature the mass number was a superscript following the chemical symbol.

$$U^{238}$$

- Computer terminology is not capitalized except in the case of specific acronyms or brand names of hardware, software, and applications.

bit	C++
byte	ADA
floppy	ASCII
program	Windows
input, output (verbs)	WordPerfect

Radiation

- The names of electromagnetic waves are usually lowercased, although *X-ray* is more commonly capitalized.

  beta ray

  gamma ray

  x-ray or X-ray

Appropriate symbols are also acceptable.

β-wave $\qquad\qquad$ γ-ray

International Units

- *Meter* and *liter* are not capitalized. The alternative spellings are *metre* and *litre*, which are used internationally. Be consistent in the alternative you select.

Quotations

- Capitalize the first word of a quoted sentence introduced by a tag (*she said, they announced, he explained*).

    The biologist explained, "The 1972 ban on DDT was an important first step in environmental responsibility."

- Do *not* capitalize the first word of an integrated quote that is part of the logical structure of a sentence.

    Gas lighting has a humble origin in the research of the British rector John Clayton, who distilled coal and "observed that the spirit which issued out caught fire."

Brand Names

- Brand or trade names are capitalized.

    Adobe

    Excel

    Macintosh

    Xerox

    Do *not* add the symbols indicating registered trademarks: ® and ™

Exercises Chapter 5, Section B

Correct the errors in the following sentences.

1. At the conference, professor Feynman proposed the theory.

2. The Doctor arrived just before the operation.

3. There was a conference for hispanic engineers.

4. The members of the American association of the academy of sciences enthusiastically await the nobel prizes.

5. He was a Nineteenth-Century french catholic, but he supported the copernican theory of Heliocentric Astronomy.

6. Einstein's unified-field theory is presented in the book the meaning of relativity.

7. "Strep Throat" is a form of Pharyngitis.

8. *The pitcher had an mri to determine whether he had a tear in the deltoid muscle.*

_____

9. *The Epidermis is considered an organ.*

_____

10. *The species in the phylum dinoflagellata cause red tide.*

_____

11. *Scorpion stings can be dangerous, and a sting from one variety—centruroides sculpturatus—calls for immediate medical attention.*

_____

12. *Sulfuric Acid is largely used for fertilizer.*

_____

13. *A german physicist developed the first practical method of producing Ammonia.*

_____

14. *He uses the Grammar Checker in word to correct his errors.*

_____

15. *She used adobe products for most of her projects, and then she used visio for the drawings and Cad for the schematics.*

_____

Core Errors: The Pronouns

This chapter identifies common and frequent writing errors. The errors concern words, of course, but, apart from a misspelling, a word is not usually an "error" as such. "Ain't" and "irregardless" and similar words are errors, but they are easily identified. Most of the serious difficulties that arise concern *misuse;* what is done with a word or phrase is the problem. Misuse is usually determined by the sentence. The error is generally a faulty *application* of a word in a particular context. For this reason, the following discussion will overlap several of the preceding chapters.

```
         ╲  Sentence Logic    ╱
          ╲ Sentence Symbols ╱
         SENTENCE CONVENTIONS
            WORD RULES
            Number Rules
              ╲  ╱
               ╲╱
```

The first section of this chapter looks at those errors that concern the core of the sentence, the nouns (or pronouns) and the verbs. Early on in the *Writer's Handbook*, I described the nouns and verbs as the "engines." They function as the subject and the action of every thought that is structured on paper. Without them there is no language. It would follow that these two word groups are the most elaborate and complicated of all the word families. In some sense, this is true—but with one unique twist. Yes, the noun family is probably the most extensive word group, and the verbs are certainly the most complicated; however, nouns present very little difficulty for a writer. Instead, it is the noun substitutes, the pronouns, that cause difficulties. Though perhaps not equal to verbs in complexity or to nouns in number, the pronouns cause frequent grammar collisions because of their role as stand-in nouns.

Pronoun Proxy Problems

There is probably no need for a single pronoun in the language. The language could be spoken without them. Pronouns are simply convenience items that allow a speaker or writer to clip what would otherwise be slow and tedious business. For example, I could write,

> The hydrochloric acid is placed in the beaker. The hydrochloric acid is diluted to a 10% solution. The hydrochloric acid is then used to spray the metal.

It is fairly apparent that a reader will prefer the speed and effectiveness of the pronouns.

> The hydrochloric acid is placed in the beaker. The acid is diluted to a 10% solution. It is then used to spray the metal.

The convenience of pronouns makes their role in our language an important one, a role that extends far beyond their modest numbers.

The problem is that writers easily fumble in the substitution process. The wrong pronoun shows up from time to time, and the error can be difficult to detect.

> The tags are then permanently connected to the left ears of the animals before they are sprayed.

Notice that the word *they* could refer to the animals or to the ears or even to the tags! Different readers might spray any of the three. I created this error when I was constructing a sample for this text, and I am supposed to know better.

- Pronouns are basically generic variants for nouns. They represent shortcuts.

    Dmitri Ivanovich Mendeleyev = he

Pronouns have to agree with their intended match word, called an *antecedent*. This match word may not always precede the pronoun, which can add to the confusion.

    Sagan was a gifted astronomer, and **he** had the ability to convey the enchantment of science.

    Although **he** was not aware of any precedent for the concept, **Mendeleyev** had a brilliant idea for the design of the periodic table of the elements.

As in the case of many rules we explore, no one is likely to make the *obvious* error.

Ø   A **researcher** must explore every avenue of possibility since **they** are liable for the legal outcomes of the product once it is marketed.

The preceding error is an obvious situation of impossible logic that says, more or less,

    The researcher themselves must explore . . . .

The problem is that the obvious error is *never* the problem. The subtle mistake is the one that is difficult to track down.

Indefinite Pronoun Agreements

- Most of the indefinite pronouns take singular verbs. They are generic references to persons and objects, but their singular or plural status is not always clear. By common practice, many take singular verbs even though this may seem a little odd, since three of the major pronouns in this group begin with the word *every,* but the pronouns mean *every single one.*

everybody   **everyone**   **everything**

Remember, the preceding three pronouns *are singular.* Other such pronouns include

anybody	**anyone**	**none**
no one	**someone**	**something**

This second cluster has more of a sense of being singular. Two other members of the group—*either* and *neither*—are clearly singular. The application is simple enough.

> **Everybody** is on the job.
>
> Is **anyone** home?
>
> **Neither** has answered.

- In the case of several indefinite pronouns (*all, any, some*) the pronoun is assumed to match the noun it represents.

> **Some** of the stars **are** red giants.
>
> **Some** of the mercury **is** lost in vapor during the life of the amalgam.

Relative Pronoun Agreement

- When you construct subordinate clauses with the relative pronouns *who, which,* and *that,* the verb of choice agrees with the noun to which the pronoun refers (the antecedent). This sounds complicated, but it is no different from the way you use the more obvious pronouns.

In other words, pronouns generally agree with the nouns they describe.

> **he** = John = **(one)**
>
> **they** = John and Sue = **(more than one)**

The same logic applies to *who, which,* and *that,* except you have to find the noun to figure out the number (singular or plural). In other words, you can't tell, at first, whether *who, which,* and *that* are singular or plural.

> **John** is the employee **who is** going to drive
>
> **Sara and Dale, who are** coming along, will be there, too.
>
> The great solution to bacterial infection was **antibiotics, which cure** a host of deadly infections.
>
> The problem was **dysentery, which causes** 50% of all fatalities in warfare.

Personal Pronoun Agreements

Although no one would ever confuse the singular or plural forms of pronouns (*he, they*) or the gender of pronouns (*he, she*), this small group of pronouns creates problems because of the different roles they play in our speech patterns. Essentially, the difficulty is that there are three variations for each of the types. The choice—of *I, me,* or *my,* for example—depends on the function of the word in the sentence.

subjective case	objective case	possessive case
(functions as a subject of a verb)	(functions as an object of the verb, an object of a preposition, or an indirect object)	(functions as a possessive and becomes an adjective or can be a noun)
I	me	my
you	you	your, yours
we	us	our, ours
he, she, it	him, her, it	his, her, hers, its
they	them	their, theirs

Examples of the various applications follow.

- *Subjective*

    She and Ruth went out for more samples.

- *Objective* (functions as an object, direct or indirect, of the verb, or as an object of a preposition).

    I took **them** to the computer fair.

    The lab tech handed **her** the manual.

    The professor assumed that the papers were for **him.**

- *Possessive*

    It was **my** idea.

    He was fired because of **his** teachings.

    Theirs is the most expensive.

The common pronoun practices seem obvious enough, but a few constructions cause some confusion.

- The *subject complement* consists of the words that follow a linking verb to complete the logic of the subject. This construction puts subjective case pronouns after the verb, because they are linked to the subject and so they have the same case.

> The radio commentator explained that the reporters were **he** and the television anchorman.

> It **was** probably **she** who found the entrance.

As in the case of several kinds of sentences, many people simply do not use the correct forms because they sound old-fashioned or stilted.

> It was **I**, sir.

- To complicate matters further, if the pronoun is the subject of an infinitive verb form (*to* + verb), it must be in the objective case because the pronoun is not a subject in this case. This results in another frequent confusion. The sentence

> Ø   The French press expected Pierre Curry and **she** *to discover* radium

should read

> The French press expected Pierre Curry and **her** *to discover* radium.

The sentence

> Ø   The council wanted Smith and **I** *to do* all the work

should read

> The council wanted Smith and **me** *to do* all the work.

- Comparisons that are stated with *than* or *as* create another difficulty because the verb is usually missing from the second half of the construction. If the sentence is parallel in verb use, there will be no problem.

> He had more data than **she** had.

However, we often omit the second verb, particularly in speaking. The result is frequently incorrect.

> Ø   The mechanic worked faster than **them.**

In speaking, the error is hardly noticed, but in writing the pronouns must be correctly used.

> The mechanic worked faster than **they.**

Tip: If you end a *than* or *as* construction with a personal pronoun, insert the missing verb to be sure of the correct pronoun. You can always remove the verb if it sounds awkward, but you will have the correct pronoun in place.

The mechanic worked faster than **they** (worked).

- If all the correct uses sound odd or stilted to you, pay close attention to your usage because, otherwise, you will favor the popular incorrect forms.

Vague Pronoun References

- Avoid using a pronoun that refers to a potentially unclear antecedent. In other words, place a clarifying noun after the words *this, that, these, those*. For example, the sentence

    Ø   **These** were what the NASA landing team had to deliberate

should read

     These proposals were what the NASA landing team had to deliberate.

The sentence

    Ø   The technician had more trouble with **that** than with the others

should read

     The technician had more trouble with **that CPU** than with the others.

- Do not use *this, that, these,* or *those* in the first or transitional sentence of a paragraph unless you add the clarifying noun. It is a popular mistake to use these pronouns as transitions when, in fact, they may confuse the reader.

¶

¶

     **We Can clearly see
that this is the preferred shape.**

A paragraph will usually develop enough material to make a pronoun transition very confusing. It is as though you were to discuss a series of geometric shapes in one paragraph and refer to only one in the opening of the next paragraph.

Pronoun Redundancy

- Avoid the following *it* construction.

    Ø   In the book **it** states . . . .

    Ø   In the report **it** contends . . . .

    Whereas we do need a clarifying noun after some pronouns, we do not need a pronoun after a perfectly obvious noun. Let the book or the report speak for itself.

    The book states . . . .

Pronoun Drift

- Avoid the use of *they* when you mean a company or agency or other grouping.

    Ø   **Texas Instruments** produces a chip that contains one million transistors in a space 1/2 inch square. **They** produce the chips in environments 1000 times cleaner than an operating room.

    This sentence should read

    Texas Instruments produces a chip that contains a million transistors in a space 1/2 inch square. **The corporation** produces the chips in environments 1000 times cleaner than an operating room.

- Avoid references to the reader through unnecessary *you* constructions.

    Ø   **You** can see that the findings are precise.

    This sentence should read

    It is obvious the findings are precise.

- If you refer to people, use the word *who* rather than the word *that*. The sentence

    Ø   All the volunteers **that** were accepted for the study met the criteria

    should read

    All the volunteers **who** were accepted for the study met the criteria.

- Use the word *that* to refer to objects. It is easy to avoid this common error if you remind yourself that it is not polite to refer to people as things. As an example, a baby is always *he* or *she* and not *it*.

Exercises Chapter 6, Section A

Correct the errors in the following sentences.

1. Everybody on the film set in the dusty valley in Utah, including the stars John Wayne and Dick Powell, were endangered because the dirt was contaminated with cesium 137 and strontium 90 from dirty nuclear bomb explosions in the Yucca Flats in nearby Nevada.

2. Anyone on the sets of the 1956 film The Conqueror were at unusually high risk of cancer twenty years later.

3. Each of the crew members who breathed the desert dust were in potential trouble by 1975.

4. A French woman worked on the DNA research with Crick and Watson, but the men received more acclaim than her.

5. Night-flying mosquitoes are hardly visible, but the substance luciferon illuminates fireflies and makes this bioluminescent insect more visible than them.

6. *The students in the computer lab had more trouble with that than with the other software.*

_____

7. *In the report on aquifer water depletion it states that the southwestern states will have water rights conflicts if precautions are not taken.*

_____

8. *Kodak introduced the disposable camera in 1987, and they gave it an appropriate name: the Fling camera.*

_____

9. *All of the computer users that were victims of the virus experienced costly damage.*

_____

10. *It is obvious that proper use of pronouns are good policy in technical and engineering documentation.*

Core Errors: The Verbs

Subject-Verb Agreement

Ø She do okay.

Ø I does it.

Ø Basic research **activities** in alternative lines of approach to fusion **is** critical.

Ø In recent years, our **attempts** at solving our traffic problems **has** amounted to not much more than a Band-Aid on a wound.

Without the remarkable constructions in the first of the preceding pairs of sentences above, our great traditions in blues music and rhythm and blues music simply would not be the same. The mismatches we see here are a normal part of the language of the African American culture that created the blues, work songs, and other American folk music traditions. However, like the distinctive regional pronunciations of Bostonians and New Yorkers, the constructions are outside what is considered the national norm; they are the expression of artists who are speaking the language of their community rather than mainstream English.

Since few mainstream people speak this way, it is a marvel that precisely the same errors are also fairly common among mainstream authors, as you can see in the errors in the second pair of quotations. The errors are examples of what is known as faulty subject-verb agreement, a rather large class of grammar errors. I can think of no examples of this error that are socially acceptable except the peculiar ". . . *aren't I?*"

All verb forms are tightly linked to their subjects by standard practices. Since these usage conventions are locked in, most people are brought up without any knowledge of a variation on the pattern. Unless someone is reared in an isolated environment, most verb forms are taken up by habit and are perceived as logical partners for nouns: a singular noun takes a singular verb form, and a plural noun takes a plural verb form.

How then do these blunders end up in writing? The answer is a simple equation that was mentioned earlier.

> The likelihood of a verb error is directly proportional to the distance between the subject and the verb.

In other words, subjects are not always nouns, and nouns are not always subjects. In formal writing, our sentences are long enough and complex enough to confuse us.

Quite simply, authors often mistakenly link one noun to the verb although another noun is the subject.

Ø   Steroid *use* by professional and amateur **athletes result** in harmful side effects and **destroy** the clean atmosphere of sports in athletic competition.

Ø   A sparkling *set* of thirty-two healthy **teeth are** expected to last a lifetime.

Ø   A *group* of **employees are** meeting in the lounge at noon to discuss grammar problems.

The error usually occurs if the noun nearest the verb is *not* the subject, which is what we see in boldface in the examples. The subjects of these three sentences are the words *use*, *set*, and *group*, indicated in italics.

- Compound subjects are also troublesome. A verb may have more than one subject. If the subjects have been joined by *and*, they are plural. Notice the error in this example.

Ø   Use of the 12-lead EKG, more detailed lab techniques, *and* knowledge of anatomy and physiology **enhances** med-tech capability in the emergency setting.

Ø   It is fair to say that the chance of injury *and* the safety of the skier **depends** on the individual skier.

Observe that each sample contains a plural subject. The problem is that the more nouns you use the more confusing the sentence becomes when you need to decide on the verb form. Shorter constructions usually lessen the confusion.

The EKG, appropriate lab technique, *and* precise knowledge **enhance** med-tech capability.

Injury *and* safety **depend** on the skier.

However, even simple examples can be confusing. For instance,

Ø   The thrill *and* excitement **is** gone from the sport

should read

The thrill *and* excitement **are** gone from the sport.

Usually, if you deconstruct the sentence and look at the subject and verb engines, you will figure out the correct match.

- When subjects are joined by *or, nor, either . . . or,* or *neither . . . nor* the correct verb form is determined by the subject nearer the verb.

Neither Jim nor *Bob* **was** in the lab.

Either steel alloys or aluminum **alloys are** adequate for the stress procedure.

A Microsoft *Word* application or any of the *Windows* **applications are** well designed.

- Since the elements in the series are exclusive of one another, the parts of a series may be plural or singular. The verb must agree with the *nearest* of the subjects.

  If burglary, armed robbery, or assaults are on the increase, local law enforcement agencies must respond with dedication.

  The same logic is applied to a pair:

  Neither one carrot nor 25,000mg of beta carotene **compounds compose** the ideal dietary supplement of vitamin A.

- Titles of articles and books are always handled as singular objects; even though a book may be called *Marsupials,* it remains one book.

  Marsupials clarifies our misconceptions about the habits of this unique family of mammals.

  In fact, this construction can be an awkward practice, and a singular noun is usually inserted for clear meaning.

  The **book** *Marsupials* clarifies our misconceptions.

- If you have reason to translate four-function basic math into words, there are a few rules for the subject-verb agreements. There are occasions when you might say *X plus X* or *Y divided by Y.* The problem is whether to say

  X minus X **equals** . . .

  or whether to use the plural form

  Y plus Y **equal** . . . .

  If you use the verb *to equal,* the rules are a little difficult to see; if you use the verb *to be,* you will clearly see the logic.

- Singular verbs are used for subtraction statements and division statements.

  Twenty minus twenty **is** zero.

  Nine divided by three **is** three.

- Plural verbs *or* singular verbs are used for addition statements and multiplication statements.

  Thirteen plus one **is/are** fourteen.

  Six times two **is/are** twelve.

- One exception is multiplication stated without the function, which is plural.

> Six threes **are** eighteen.

Tense Errors

Tense problems are not new. Nonstandard usage has enormous staying power in the language.

> He done her wrong.

> She done him wrong.

The preceding pair of constructions are part of our great tradition in country music. They also represent a grammatical matrix in the American social fabric that is much imitated among country singers, and yet the likes of Willie Nelson or Reba McEntire are not likely to be found talking this way outside of the sound studio. The moment they are off stage they will speak using more standard forms.

The error that characterizes such constructions is the use of a verb stem from the perfect tenses but without an auxiliary verb. In other words, the correct verb forms for the traditional country music verbs in the example would be

> He did her wrong.

> She did him wrong.

These are verb errors that were just as common 150 years ago.

> Well, I reckon I got to light out for the territory . . . .

> I been there before.

Progressive tenses create the same problem:

> He doing.

The construction should be

> He is doing.

A variant effort to use the auxiliary verb usually backfired and created

> He be doing.

There is not even a remote chance that you would make tense errors of this sort. It is nice to know that you are a master of singular and plural forms of twelve tenses of a thousand verbs or more. In exponential math, this means you will make no errors in a million instances of tense applications. Pareto would be pleased. This is comforting news, except for two or three gremlins.

Subjunctive Mood

The subjunctive mood is used to express wishes, orders (requests), and conditions that are contrary to fact. The verbs in the subjunctive mood are either "were" or the base term of the verb (*be, speak, work*). You have probably seen these curious constructions often.

> She wished she were finished with her graduate exams.
>
> The committee requested that the report be revised.
>
> He recommended that she study three hours a day.
>
> If I were president of the country, I would propose much stronger pollution guidelines.

Perhaps the most popular subjunctive is the expression

> If I were . . . .

Most American speakers state these constructions in ways that avoid the subjunctive. For this reason, subjunctive construction are more likely to appear in writing.

Tense Shifts

Time changes can result in confusion for a reader. This is not a grammatical problem. The problem is strictly logical. A reader will try to follow the time frame that a writer constructs in a document.

Your logical reference for proper time-frame management is simple enough. You place the time frame in the *general past,* the *general present,* or the *general future.* Occasional comments will stray out of the time frame by using the *perfect* or *progressive* forms of the same tense. Notice the way I maintain my position in all three of the present tense forms in the following sample.

> I **study** a great deal of scientific material. I **have studied** a considerable number of articles concerning archaeology. I **am studying** books on astronomy at the moment.

In truth, you can shift from any tense to any other tense as long as the logical continuity of your thinking is maintained.

Authors can manipulate a number of tenses for each general time frame, although they use the past tense as their primary time position in most writing. They do this because writing is generally historical, even though you may not notice this writing convention.

There are exceptions, such as sets of instructions that are in the present tense, but writers rely heavily on the past tense.

You can exit the time frame at any point, although a new paragraph may be the least confusing place for the shift. a new paragraph will open with a topic sentence that is supposed to signal a new topic, and it may provide a transition to the new topic. What it can *also* do is change the time frame, even though this is seldom mentioned in style manuals.

> Once the findings of the analysis were completed, we looked at the outcomes. There **are** . . . (shift to present).

> Huxley's observations are true but he overlooks the conditions that preceded his studies. In the 1830s there **were** . . . (shift to past).

Transitions are a practical tool for assisting the reader along the paths of your material, including the time logic of your tenses (see Chapter 10 in *Basic Composition Skills*)

Literary Time

There is one literary convention you need to handle with skill. If you discuss any work of literature, regardless of the age of the text, feel free to bring the author to life. Our great thinkers are always happy to walk again among us.

> Darwin published *Origin of Species* in 1859. In the text he **explains** the famous theory of the survival of the fittest.

Notice that I did not leave the poor fellow in the dust. I discuss him as a living, breathing scientist. This wonderful convention may not seem logical, but no one seems to care. Perhaps we could use the past tense instead. You will notice, however, that the practice of present-tense discussions of the literature and the thinking of our ancestors is quite common.

Irregular Verbs

For the hundreds and hundreds of regular verbs, the past-tense and the past-participle forms are the same, meaning they end in *-ed* or *-d*.

Base Form	*Past tense*	*Past Participle*
walk	**walked**	**walked**

Unfortunately, some of the verbs that are used the most are highly irregular, even though they are far fewer in number. The term *irregular* simply means that these verbs do not follow the usual predictable pattern for verb-form endings. In this group of over one hundred verbs, there are a small number of irregular forms that are frequent problems for writers. You are probably well aware of these verbs; for example, think of the well-known problems most people have with the verbs *to lay* and *to lie*. Review the following chart of problem verbs, and expect trouble if you use them. The computer software is of little help, so you will need to carefully examine your use of any of the following verbs.

Blunder Verbs		
Base Form	**Past Tense**	**Past Participle**
awake	awoke, awaked	awoke, awaked
cling	clung	clung
dive	dived, dove	dived
dream	dreamed, dreamt	dreamed, dreamt
forget	forgot	forgotten, forgot
hang (suspend)	hung	hanged
hang (execute)	hanged	hanged
lay (put)	laid	laid
lead	led	led
lend	lent	lent
let (allow)	let	let
lie (recline)	lay	lain
prove	proved	proved, proven
ring	rang	rung
shrink	shrank	shrunk, shrunken
spin	spun	spun
strike	struck	struck, stricken
swing	swung	swung
wake	woke, waked	waked, woken
wring	wrung	wrung

Other Verb Errors

Most of the miscellaneous other verb problems are variations on the problems identified earlier. A number of the agreement problems relate to the issue of pronoun confusion. Many of the following rules reflect the earlier discussion of pronouns except that, earlier we looked at pronoun agreement with nouns. Here we examine how the verbs agree with nouns and then pronouns.

- The verb agrees with the subject and not with other nouns that come between the subject and the verb.

    Ø   Atmospheric *changes* in hot, tropical **weather creates** the ideal environment for hurricanes.

    Ø   The methane *storm* of Jupiter, like the solar flares of the sun and the volcanic moons of several **planets, are** enormously active.

Both these samples are incorrect. Writers tend to make the verb agree with the nearest noun (in bold) rather than the real subject (italics).

- Compound subjects connected by *and* generally take a plural verb.

    The **dog** *and* the **cat were** in the same cage at the clinic.

    Fingerprints (first used in 1902) *and* **autopsies** (first conducted in 1908) **were** important procedures in criminology.

    DDT *and* **Chlordane are** thought to produce xenoestrogens, which are possible triggers for cancers in women.

- Most indefinite pronouns are singular:

anybody	**anyone**	**no one**	**someone**
everyone	**everything**	**everybody**	

    Everybody is on board.

    Anyone without a life preserver is behaving irresponsibly.

- Several of the indefinite pronouns—*all, any, some*—can be singular or plural because they assume the singular or plural status of the noun (or other pronoun) to which they refer.

    Some of the *infection* **was** removed.

    All the fire resistance *specifications* **were** provided by the ASTM requirements.

- The relative pronouns *who, which,* and *that* agree with the nouns (or other pronouns) to which they refer. They can be singular or plural, and the verb form must correspond.

    The *gears* **that are** ready can be installed.

    The damaged *cog* **that is** on the old gear assembly should be saved as an example of stress damage.

    The doctor is the *provider* **who files** the insurance claim.

    The therapists are not the *providers* **who file** the insurance claim.

- Some apparent plurals are actually singular, and a few apparent singulars are actually plural. Fortunately, the confusion is limited to a handful of words. A number of these nouns look plural because they end in *s* but are understood as singular in meaning:

  electronics　　**mathematics**　　**physics**　　**economics**　　**news**

  They use singular verbs:

    Mathematics is great fun.

  A very few nouns have somewhat the opposite logic.

    L. L. Bean used to sell an outdoor product called a "pant." This was probably a pair
    of pants cut in half with a scissor.

  Neither word, *pant* or *scissor,* has a singular form. The plural form (*pants, scissors*) is understood to be singular or plural.

- Collective nouns act as singular subjects. A wide variety of words define groupings; these include *team, squad, battalion,* and a number of other groups in sports, civil agencies, military units, and similar groupings, as well as general words such as *crowd, audience, and committee. Data* is usually used as a collective noun also.

    The squad advances, and the crowd moves away.

- A linking verb agrees with the subject and not its complement. We sometimes mention the subject on *both* sides of a verb (dogs are canines), and sometimes one is singular and one is plural. That is, the verb agrees with the noun before the verb.

    Viruses were the subject of this lecture.

    The **subject** of his lecture **was** viruses.

    Her **research is** all the diagnostic problems.

    All the diagnostic **problems are** her research.

Lie and Lay

- Use the synonyms for *lie* and *lay. To lie* means to recline or to rest on a surface. It is easy to select one of the synonyms to avoid the grammar problems.

    to sleep, to nap, to rest

To lay means *to place,* and the synonym *to put* is a popular choice of an alternative.

- Apart from confusion about the meanings of the two verbs, however, any stab at a conjugation is often wrong also. The correct forms need to be learned.

Base Form	Past Tense	Past Participle	Present Participle
lie	lay	lain	lying
lay	laid	laid	laying

Active Voice

In the active voice, the subject of the sentence performs the action. Active voice means there is an active subject.

> Highway engineers developed the cloverleaf ramp systems.

In the passive voice, the subject *receives* the action. The performer of the action is repositioned to come after the verb, often after the preposition *by.* Passive voice means a passive subject (see also pp. 98–99 and 117–119).

> The cloverleaf ramp systems were designed **by** highway engineers.

The active voice is popularly preferred in writers' guides because the sentence construction is less complex, the verb tenses are less complex, and the subject is in the frontal position, where it is easily recognized. A passive construction such as

> Projects are initiated **by** the R&D team

is easily reversed to read

> The R&D **team** initiates the project.

The passive construction often lacks the *performer* of the action of a verb:

> The defect was analyzed promptly.

In this case a subject that is also the performer can be placed in the sentence.

> **She** promptly analyzed the defect.

In technical writing, passive constructions are generally discouraged, although there are applications in which the passive voice is a practical tool. In directives, instructions, manuals, and other documents that provide instruction, the passive voice may be used for a number of reasons:

- To focus on the *task* or the *intent* of instruction:

    The delicate diamond needle is very gently slipped into the mounting.

- As an *alternative* to the imperative mood:

    Only a reciprocating saw is used for this cut.

- To avoid frequent references to "you."

    The bolt is threaded through the housing. The gasket is fitted to the bolt. The washer is placed against the gasket. The nut is torqued to the specifications.

As long as the construction is quite clear, the passive is effective. As I noted earlier, however, you may be asked to avoid passive constructions.

Exercises Chapter 6, Section B

Correct the errors in the following sentences.

1. There is a number of complex chemical reactions in smog.

2. When you look at them, the early Ford car and the original Ford truck always has the same appearance because of intentional engineering design continuity.

3. The replaceable parts, the consistent design, and the invention of the moving assembly line was an innovative contribution that Ford made to manufacturing technology.

4. Some people thought that the mystery and glory of astronomy was gone after we landed on the moon.

5. Neither the pulsar nor the quasar are clearly understood except as theoretical models.

6. Precious Stones were a hot item in the book stall at the gem show.

7. Nine hundred samples minus two hundred samples leave seven hundred samples.

_____

8. Seven twos is fourteen.

_____

9. B. F. Goodrich introduced tubeless tires in 1947, and in 1948 Michelin introduces the radial tire.

_____

10. The possibilities of risk from radon contamination increases if the environment is not well ventilated.

_____

11. Low frequency electromagnetic fields (ELF) and the electromagnetic radiation (EMR) from domestic devices, such as the toaster or the hair dryer, radiates low levels of energy that might pose a health risk.

_____

12. Light alloys was the secret to fuselage design for aircraft.

_____

13. The focus of much of Van de Graaff's research at Princeton were methods for generating extremely high voltages.

_____

14. The first electric traffic light with green and red signals was installed at the corner of Euclid Avenue and 105th Street on August 5, 1914, by the city engineering department of Cleveland, Ohio.

_____

15. The paperclip, a well-established product of the twentieth century, was invented in 1899 by the German Johan Vaaler, who was granted a U.S. patent two years later.

_____

16. The expression "World Wide Web" was coined at CERN (home of the European particle physics lab) by Tim Berners-Lee in 1980.

_____

17. The HTML programming language that is used on the Web was also developed by Berners-Lee.

_____

Sentence Component Errors

Sentence Completeness

The fact that the language is spoken in sentence fragments somewhat explains the tendency for authors to use them in writing. Also, because punctuation is generally not spoken, except for questions and exclamations, writers may create two similar problems, called *splicing* and *fusing*. Since punctuation is voiced with pauses and inflections (if it is done so at all), the logical unit of a sentence and where it stops is not always obvious. In writing, these errors are critical problems. A reader can wade through minor errors, but a reader should not have to make an effort to understand sentence constructions.

Sentence Fragments

Sentence fragments usually occur whenever either the subject or the verb is missing. This may seem glaringly obvious in a short sentence.

> She runs well.
>
> . . . runs well.
>
> She . . . well.

However, the usual error encountered in formal writing is a clause that should have been linked to a parent sentence that preceded it.

> She runs well.
>
> Which is testimony to effective training.

The second clause is what is called a *dependent clause;* it depends on the base sentence or parent sentence to make sense. By itself it is incomplete in logical content. Since you know a fragment when you see one, be alert to this particular dilemma—the subordinate clause fragment—which is an occasional problem even for skilled writers. A sentence will be suspicious if it begins with such words as

after	**unless**
because	**when**
before	**where**
if	**which**
though	

These words can signal a perfectly healthy introductory phrase for a sentence. If they are fragments, they are the tail on the dog; any one of these words may be initiating a clause that is supposed to be attached to the preceding sentence.

> I explored the cave for several hours.

> Ø   **Because** I was looking for crystal formations.

Simply join the fragment to the first sentence, since the fragment completes the logic of the base unit.

> I explored the cave for several hours because I was looking for crystal formations.

Comma Splices

Comma splices occur when sentences are joined with a comma. Spliced sentences are called *run-on sentences*. A comma alone cannot be used to connect sentences.

> Ø   I was late, John was on time.

Remember that the comma is not adequate for joining sentences. You must support the comma with a conjunction:

,and   **,but**   **,or**

Then the sentence will be structured correctly.

> I was late, **but** John was on time.

Fused Sentences

Fused sentences occur when two sentences are run together without punctuation. This is admittedly rare among college-educated writers and usually signals a typographical error:

> Ø   I prefer Macintosh systems Ana prefers IBM-PCs.

Two rules will help resolve these difficulties in sentence construction. We observed earlier that coordination and subordination are drivers in the logical design of sentences. Two common constructions have evolved out of that logical need (see also pp. 21–22).

1. Use simple conjunctions—*and, or, but, for, nor, yet, so*—to join sentences that you want to coordinate, and use a comma with the conjunction.

> Japan has more than a dozen Nobel laureates, **but** American scientists have received over 200 of the awards.

2. Use the larger connectors (conjunctive adverbs) to link sentences in which the logic link is somewhat more complex or more precise:

however   therefore   nonetheless   furthermore

Be sure to place a semicolon before the connector *and* add a comma after the word. This is a three-part structure.

;however,

The sentence is then structured with this standard device:

> The placement of the windows seemed high; **nonetheless,** they conformed to the engineer's interior elevation drawings.

Misplaced and Dangling Modifiers

Modifiers are intended to describe. The description consists of a *word,* a *phrase,* or an entire *clause.* Writers are more likely to misplace longer modifiers because of a camouflage effect that every writer encounters. There can be a great deal of activity going on in a sentence; logical errors can be well concealed in long sentences. One of the older grammar books uses this well-known sentence as an example of what not to write.

> Running along the street, his nose was cold.

I still use it for a chuckle, but it lacks the subtlety that normally leads to the problem of dangling and misplaced modifiers. There are a number of variations on this error, and they are not always obvious:

Ø   The pilot positioned the infrared camera while the plane held steady to shoot hundreds of frames of film.   (misplaced)

Ø   Before tackling the program, the software was replaced.   (dangling)

Notice that the person who was working on the software is nowhere to be found in the sentence. And the first sample? Planes do not shoot film, of course, so the sentence is not exactly correct.

> The pilot positioned the infrared camera to shoot hundreds of frames of film while the plane held steady.

Misplaced Modifiers

Keep any modifying phrase close to the word it modifies. This sample is a little more obvious than the two preceding ones.

> Ø   Although weighing over a thousand pounds apiece, the engineers chose prestressed hollow core slabs for the twenty-foot span.

This sentence means

> The engineers chose prestressed hollow core slabs, although weighing over a thousand pounds apiece, for the twenty-foot span.

Here is another example:

> Ø   Even though they are only 1/50 the width of a human hair, semiconductor engineers measure chip tolerances of semiconductors in microns.

This sentence should read

> Semiconductor engineers measure chip tolerances of semiconductors, even though they are only 1/50 the width of a human hair, in microns.

The term *mismatched modifier* is popularly used to describe modifiers on the order of planes that shoot film, but the mismatch can be subtle if the modifier almost fits the logic in the sentence.

> Ø   Although only a small percentage in error, the researcher decided the initial outcomes were the major problems in the study.

> Ø   The researcher decided the initial outcomes were the major problem in the study, although only a small percentage in error.

The problem is that neither the researcher nor the study is the real source of the errors—which were in the initial outcomes. We have to move the modifying phrase to the middle of the sentence to position it near the word it describes.

> The researcher decided the initial **outcomes, although only a small percentage in error,** were the major problem in the study.

Too subtle? Yes, the mismatched modifier can be difficult to find at times. Of course, if no one sees the problem with the phrasing, and no one is confused by it, then the error is not critical. If the errors are obvious, however, they are very obvious.

Dangling Modifiers

In a worst-case situation, there is no subject in the main clause to receive the description of an introductory modifier. The following samples show how increasingly subtle the problem can become:

Ø   When discovered in shallow water, their playfulness was surprising.

(We don't know what was discovered.)

Ø   Being unstable, its effectiveness was inadequate.

(We don't know what was unstable.)

Ø   To gain acceptance by the natives, trinkets were usually distributed among them.

(We don't know who wanted to gain acceptance.)

Ø   When viewed from this perspective, there was no doubt about the decision.

(Who or what was viewed from this perspective?)

Ø   Though quickly responding to the symptoms with an antibiotic, the pathogen was nonetheless spreading.

(Who quickly responded and provided the antibiotic?)

Ø   Having installed the cables, the network was ready to test.

(Who installed the cables?)

The dangling modifier always comes at the beginning of a sentence. As you can see, a very common problem that causes a dangling modifier is the simple absence of the word that the modifier is supposed to describe. That noun should come after the opening modifier. This is a particularly common logical problem if human processes are confused with or assigned to mechanisms we invent.

> Instead of measuring the strategic risk, the plane fired a smart-system missile at the engine of the train.

First, unless the firing was an automatic function, the pilot fired the missile. Second, strategic risk is not programmable. The modifier is talking about the pilot, not the plane.

> Instead of measuring the strategic risk, the pilot fired a smart-system missile at the engine of the train.

The following samples are a few of the alternatives that can be constructed to correctly use the modifiers in the preceding group of sentences.

When discovered in shallow water, the *seals* demonstrated their playfulness.

Being unstable, the *antibiotic* was ineffective.

To gain acceptance by the natives, *traders* distributed trinkets among them.

When viewed from their perspective, the *proposal* was a logical one, and the members of the committee had no doubt about the decision.

Though quickly responding with an antibiotic, the *doctor* realized the pathogen was spreading.

Having installed the cables, the *technician* was ready to test the network.

Exercises Chapter 6, Section C

Correct the errors in the following sentences.

1. Xerox developed Ethernet. Which is an access protocol designed for local area networks that are set up with a ring structure.

2. The pull-tab on today's beverage can was introduced in about 1980. Although the patent design for this humble outcome of industrial engineering was issued seven years earlier.

3. Von Braun and 120 other rocket engineers surrendered to the Americans at the end of World War II. And Von Braun later led the design team for the Saturn rocket that placed the American astronauts on the moon.

4. Nitrogen is abundant and essential for life. But the triple-bond structure of the molecule was a challenge to chemists who sought to develop synthetic fertilizers.

5. Rejecting constant design changes, the early Ford automobiles were consistently manufactured to look the same.

6. To create new market drive and purchasing interest in cars, frequent change became an annual event for all GM cars.

_____

7. With an unmistakable new silhouette, Douglas Aircraft produced over 10,000 of the legendary DC-3 planes.

_____

8. Having evolved to look toxic, the theory of Batesian mimicry explained one reason why some animals will look like other animals.

_____

9. Greenhouse gases include a number of synthetic pollutants that are products of manufacturing, which contribute to global warming.

_____

10. When considering the risk of nuclear war, the dangers of weapons manufacturing know-how is obvious.

_____

11. Having introduced a steam-driven press in 1814, a thousand pages of the London Times could be printed every hour.

_____

Other Common Errors

Adverbs

Adverbs are used to modify verbs, adverbs, and adjectives. It is fairly common to hear adjectives where there should be adverbs—which usually end in *ly*. The error is less common in writing. Remember that most adverbs are adjectives to which the -ly ending is added (although there are exceptions.)

Ø The test results looked **awful** good.

The test results looked **awfully** good.

Ø The acoustics in the chamber worked **perfect.**

The acoustics in the chamber worked **perfectly.**

Gerunds

Almost any verb can be used in reference *to* its activity rather than *as* its activity. We attach the *-ing* ending and *run* becomes a noun, an image of an action that is ongoing:

I am **running** (active verb)

I enjoy **running.** (gerund)

These verbal nouns are often used with a possessive noun or pronoun.

The **snake's hissing** was an unmistakable warning to the geologists.

The **doctor's probing** seemed to lead to another case of iatrogenic disease.

Comparatives and Superlatives

Adjectives and adverbs usually have three forms depending on the number of objects they are describing. They have a simple form to describe one object.

The nose cone was **dull.**

A second form is used to compare *two* things.

The Shooting Star nose cone was **duller** than the nose cones on most of the planes.

A third form is used to compare *three or more* things.

The **dullest** nose cone design in the flight museum collection of jets was created for the old MIG-15.

The process of forming comparatives and superlatives for monosyllabic adjectives (or syllables) consists of adding *er* and *est*.

	Comparatives	Superlatives
fat	fatter	fattest
thin	thinner	thinnest
old	older	oldest
young	younger	youngest
slow	slower	slowest
hot	hotter	hottest
cold	colder	coldest
big	bigger	biggest
small	smaller	smallest

A few of the most popular constructions are irregular.

good	better	best
bad	worse	worst
little	less	least

Less and *least* refer to volume or quantity, and *smaller* and *smallest* refer to size. Many comparative and superlative adjectives can also be formed by adding the words *more* and *most*. Polysyllabic adjectives (those consisting of more than one syllable) form their comparatives and superlatives by adding the words *more* or *most*.

generous	more generous	most generous

He was **generous.**

She was **more generous** than her brother.

She was the **most generous** of the four sisters.

The logic can be structured in reverse by using the words *less* or *least*.

appropriate	less appropriate	least appropriate

The Gremlins

There are some areas of writing-structure problems in which flexibility will win the day. For example, if you do not know the *lie* and *lay* distinctions, you simply use another verb, such as *rest* or *put*. The path of least resistance should probably be one of the guiding principles of a writer. In this gremlin section, you will find a small group of rules and practices that you more or less cannot avoid. You will have to check these rules from time to time.

- **a, an**

  We generally use the article *a* for words that begin with consonants and *an* for words that begin with vowels.

a damper	**an** element
a condenser	**an** assembly
a footing	**an** annunciator
a circuit	**an** exhaust
a baffle	**an** evaporator

  Because we use a great many abbreviations in the technical disciplines, a few problems emerge. Most abbreviations and acronyms will be consistent with the preceding general rule.

a RAM	an EPA decision
a LAN	an ISA workstation
a CPU	an ECI workstation

  Exceptions occur because one group of letters, when pronounced, sound as of they start with a vowel. For example, look at the use of *fine* and *F* in this sentence:

  A fine lab project would not get **an F.**

  Certain consonants, like *F,* have a vowel sound. Here are the approximate pronunciations.

Sound	Letter	Sound	Letter
ef	(F)	are	(R)
el	(L)	es	(S)
em	(M)	eks	(X)
en	(N)		

They are matched with the appropriate article by pronunciation, as in *an F grade*. The result is that abbreviations beginning with *F, L, M, N, R, S,* and *X* use the article *an*.

an SLR   **an** X-ray   **an** M-16

A much more complicated problem arises with the selection of an article for an abbreviation when the abbreviation is never spoken in the abbreviated form. This is an uncommon situation, but it can be sticky.

Chemists tell me that although they write in chemical formulas, they are "not supposed to speak the formulas" (although they do). In other words, a chemist will write CO_2 but say "carbon dioxide." Fortunately, the elements and their compounds are mass nouns (see p. 000), and so they are usually accompanied by the article *the*, since they are thought of in volume rather than in numbers. In other words, we say

the sulfuric acid   (the volume)

the H_2SO_4 (the volume)

If we refer to a single molecule, we choose the article preceding the element to accommodate the pronunciation of the element name.

In mass spectroscopy, **a** SF_2 molecule will fragment differently than **a** CO_2 molecule, and **a** NO_2 molecule will fragment differently, also.

a sulfur difluoride molecule

a carbon dioxide molecule

a nitrogen dioxide molecule

There are several solutions to these occasional conflicts:

• Avoid abbreviations in the running text if no numbers are involved.

• Avoid abbreviations if there is a conflict between the abbreviation and the way it is spoken.

• Shift to the definite article *the* if the shift from the indefinite article *a* is acceptable in terms of the logic of the sentence. In digital electronics, *a* flip flop can be abbreviated as *an* FF, but it would not be called "an FF," so avoid the clumsy problem by writing it as "*the* FF." The SCSI drive is a similar issue.

- **amount, number**

  This is a word-choice problem and not a spelling problem.

  Use the words *amount of* in reference to mass nouns that are measured in volume rather than number. Examples include *cement, electricity, gold, oxygen, water, steel.* A mass noun is obvious because it lacks the plural *s*. In general, this group of nouns includes gases, liquids, and the elements of the periodic table (see also pp. 217–218).

  > There was a small **amount of water** in the fuel storage tanks.

  Use the words *number of* (in the sense of *many*) in reference to things that can be counted out: *trees, test tubes, clocks, cows, comets.* If the plural form has an *s*, the items can be counted.

  > There were a **number of** storage **tanks.**

- **and/or**

  Avoid this construction. It creates confusion, even though it attempts to create precision. The construction forces the readers to state the options for themselves. Use the word *and* or the word *or* if appropriate. If the sentence genuinely means to indicate both options, recompose the sentence so that the options are correctly stated. The author must do the work, *not* the reader.

  > The lab technician removes the stains with cleanser **or** bleach, **although both may be needed.**

- **ampersand**

  The symbol & is not used in running text. Use the ampersand only as part of an organization name, usually a company, and only in footnotes, bibliographies, and other types of lists that are set apart from the text of the document.

- **as**

  Avoid this substitute for the word *because.*

  > Ø   **As** the leg was fractured, it was difficult to remove the driver from the car.

- **can not**

  The word is *cannot.* The two words were fused long ago. Nonetheless, writers will still incorrectly separate the words, just as they will do with *nonetheless.*

- **datum, data**

  Datum is singular.

  > The simplest **datum** is used to help the investigation.

  Data is plural and is the usual word you will want.

  > The **data** are easy to compile.

  The word is plural, but it is now so widely used as singular that the correct form may sound awkward to many readers. I suppose you have to please the greater population of your readership. If you say,

  > The data has been gathered

  there will be hardly a yawn among your readers, but the true plural form is

  > The data have been gathered.

  This is a judgment you have to make, since the correct usage might startle your readers.

- **fewer, less**

  The word *fewer* is used with nouns that can be counted, which are usually identifiable because their plurals are formed by attaching *s* to the nouns.

  > fewer logs   fewer flowers   fewer pipes

  The word *less* is used with mass nouns that are measured in volume. These nouns usually include gases, liquids, metals, and the elements of the periodic table. They are identifiable because they do not form plurals with *s* (see also pp. 217–218).

  > less air   less gold   less sulfur

- **good, well**

  Good is an adjective and describes a noun.

  > We had **good** news.

  Well is an adverb and describes a verb.

  > I ran **well** in the race.

The incorrect choice is usually apparent

Ø   I ran good

except for the popular expression

>I feel good.

- **he/she**

In situations that are *clearly* generic, avoid *he*. One set of alternative constructions includes the popular *he/she* or *he-she* or *s/he*. There is no handy singular pronoun that is gender neutral, as I observed elsewhere. The word *one* is perfect for the job, but it is considered to be old-fashioned usage.

Two alternatives to these singular pronouns are practical solutions to awkward phrasing: use nouns or plural pronouns (see also pp. 44–46.) The sentence

>**He/she** welds the plate to the beam

could read

>The **welder** welds the plate to the beam.

Or, as an alternative, pronouns can be used:

>**They** must complete the research on the tenth.

- **I**

Use the *I* style for speedwriting. The first-person personal pronoun is discouraged in formal writing, as I have noted. However, one obvious trick for rapidly developing a text is to write as you speak—and just as quickly.

A number of potholes slow down a writer, including slow mechanical skills such as typing, too much self-consciousness of rules and spelling, and a stiff writing style. As many writers complain, "I erase more than I type." If this sounds familiar, try to write as you speak—with the *I*. You will be more comfortable, and you will speed up. The *I*'s are easily removed farther along the path of your project. Write with the *I*. Rewrite without it.

- **its, it's**

Not all possessives take the apostrophe. *Its* does not. Possessive pronouns also do not take an apostrophe (his, hers, theirs, yours, ours, whose).

The salamander lost **its** tail but quickly generated another.

The construction *it's* is a contraction.

    It's going to rain.

> **TIP:** Most formal writing is done with few contractions; as a result, the odds that you need the apostrophe in *its* should be remote. Do not use the contraction.

- **like, as**

Like is a hugely popular catchall word of recent years. Apart from being as annoying as the expression *you know,* the word *like* is troublesome when an author moves it into the capacity of a conjunction while writing.

The word *like* is not a substitute for *as, as if,* or *as though.* The sentence

Ø   It appeared **like** the comet exploded

should read

    It appeared **as though** the comet exploded.

In another example,

Ø   The circuit performed **like** there was a ground fault

should read

    The circuit performed **as if** there were a ground fault.

- **loose, lose**

Pronunciation is the key to remembering the distinctions.

    A loose goose could scare a moose.

Loose is an adjective.

Lose is a verb, as in

    Did the Cowboys **lose** the game?

- **sit, set**

To sit means *to be seated.*

Present	Past	Past participle	Present participle
sit	sat	sat	sitting

To set means *to place* or *to put*.

set	set	set	setting

Set is usually a transitive verb, meaning its forms will *place* or *put* something.

sit at the table	(intransitive)
set the table	(transitive)

As we noted elsewhere, most speakers diligently avoid these road hazards, particularly the word *set* in its past forms. *Put* is the typical substitute. If you use either of these verbs in your writing tasks, check the usage.

- ## so, such, very

Avoid using *so* and *such* as intensifiers to amplify the meaning of other words. The problem is that the measure of the intensity is not precise.

Ø   The French artillery officer Amédée Mannheim was **so** inventive.

Ø   His slide rule was **such** a great tool.

Of course, the word *very* is not much better, but of the three words, the most acceptable of these vague amplifiers is *very*. The word will function adequately if you add another sentence to explain precisely what you mean.

We had **very** good news. The company earned $500,000 in the first quarter.

Essentially, something *very* large or *very* small must be precisely explained. No amount of general talk will explain that something very large is *humongous* or that something very small is *eensy, weensy, beensy, teensy,* or that so much of something is *oodles,* or that something so small is *itty-bitty* or *itsy-bitsy.* The intensifiers are poor measures of our meaning, just as poor as some of the comical adjectives.

- ## the

Be sure to include the word *the* in every location that calls for it. You are not economizing by turning instructions into telegrams. Western Union messaging does not belong in the age of fax or in the Internet era. The sentence

Ø   Use torque wrench to reassemble hub by tightening nuts on escutcheon

should read

> Use the torque wrench to reassemble the hub by tightening the nuts on the escutcheon.

Inexperienced writers are inclined to mistake the confusing cryptic technique for good technical writing. Cryptic writing was once used for writing telegrams. No one speaks with the telegraphic style except those old tin robots in 1950s sci-fi movies.

- **there, their, they're**

  There is a location.

  Their signifies ownership.

  They're is a contraction (*they are*), and this form is not used in formal documentation, in which the spelled-out words are preferred. If you tend to misspell *their* as *thier,* remember that all three forms begin with the word *the.*

- **which, that**

  That is used in a restrictive clause (one that is critical to the meaning of a sentence).

  > Streets are seldom paved with bricks **that** are bumpy enough to effectively slow down an automobile.

  Which is used in a nonrestrictive clause (one that is not critical or is, in fact, clearly secondary).

  > The street department thought the brick streets would slow down the traffic, **which** remains a problem anyway.

  If in doubt, remember that a comma generally precedes a clause starting with *which,* and a clause starting with *that* is never preceded by a comma.

Exercises Chapter 6, Section D

Correct the errors in the following sentences.

---

1. It was a rare day when any of the experimental prototypes worked perfect.

---

2. Put the mouse over the arrow until a selection window shows an features list.

---

3. There was a small amount of trees that were damaged by the wind.

---

4. Chlorofluorocarbons (CFCs) include refrigerants, aerosol propellants, and/or solvents.

---

5. Less diseases now threaten human beings thanks to antibiotics and vaccines.

---

6. The buyer should compare the handheld PCs and Notebooks before he/she makes a decision to invest in a small computer.

---

7. At it's worst, a hurricane's wind velocity can approach 200 miles per hour.

_____

8. All the books about tropical illnesses will have everybody thinking these exotic diseases are on the lose.

_____

9. Thier are 800 tornadoes in the United States in an average year.

_____

10. Municipal wastes that includes all liquid discharges account for 22% of the oil pollution in the oceans.

_____

11. By using a roll of celluloid film which had a sequence of coordinated pictures on it, Edison created an early motion picture system that heralded a new technology.

_____

12. Assemble contact base, contact, and spring. Insert contact plate assembly into body using drawing and using actuator rod. Cut retention snaps.

_____

13. As you are seeing all the errors in the samples, I would say you are doing good.

_____

Inappropriate Expressions

A number of popular expressions are nonstandard. In an effort to sound formal, an author may use a variety of phrases that are unnecessary, confusing, or clumsy. These terms can find their way into many varieties of documentation.

Ø **all-right**

Ø **alright**

The one-word versions are not correct. The correct spelling is *all right*.

> The answers were **all right.**

Ø **all ready**

This expression does not mean *already,* which means *prior to* or *previously.*

> The task was **already** completed.

Ø **alot**

The correct phrase is composed of *two* words: *a lot* of . . . .

> There were **a lot** of tests.

Ø **as regards**

Acceptable but awkward. Change to **regarding** or restate the sentence without the construction.

> **Regarding** the findings, the research team was satisfied.

Ø **as such**

Remove from the text.

> The work ~~as such~~ was adequate.

Ø **as to whether**

Ø **as to when**

Ø **as to where**

Remove from the text and restate.

> Ø   They were confused **as to where** to relocate the sewage plant.
>
> They were confused about **where** to relocate the sewage plant.

Ø **cannot help but**

Ø **could not help but**

Remove from the text and restate.

> Ø   He **could not help but** defend his theory.
>
> He **had** no choice but to defend his theory.

Ø **center around**

Change to *center in* or *center on*.

> This particular behavior pattern is **centered in** the frontal lobe.

Ø **close up**

Redundant idiom. The word is *close*.

> Ø   We tried to get **close up** to take the picture.

Acceptable as photography jargon, but hyphenated.

> It was a sensational **close-up.**

Ø **different than**

Change to *different from*.

> The fossil in the gorge was **different from** the fossil on the hilltop.

Ø **each and every**

Redundancy. The word is *each*.

Ø e.g.

Use the English equivalent *for example*.

Ø i.e.

Use the English expression *that is*.

> Computer-simulated airstreams updated the former air foil observation methods; **that is,** the graphic vector velocity fields are much more revealing in the newer computer simulations.

Ø incidently

Change to *incidentally*.

Ø inside of

Change to *inside*.

> The flaw was **inside** the CPU.

Ø nowhere near

Remove from the text and restate.

> Ø   We had **nowhere near** enough light for the low ASA film we wanted to use.
>
> We **did not have enough** light for the low ASA film.

Ø on the grounds that

Change to the word *because*.

> He was convicted **because** he was involved.

Ø outside of (location)

Change to *outside*.

> The tool was **outside** the box.

Ø outside of (except for)

Remove from the text and restate.

> Ø   **Outside of** a dog, a book is a man's best friend; inside of a dog it is too dark to read.—Groucho Marx
>
> **Other than** a dog, a book is man's best friend.

Ø   **Outside of** the form tie assembly, there is no convenient device for containing the frame of a reinforced concrete wall.

Apart from the form tie assembly, there is no convenient device for containing the frame of a reinforced concrete wall.

Ø over with

Change to *completed, finished, accomplished,* and the like.

Ø   The research was **over with.**

The research was **completed.**

Ø reason is because

Change to *because.*

The plants would not grow **because** of the lack of light.

Or change to *the reason is that.*

The **reason is that** the plants lacked light.

Ø repeat again

Redundancy. The word is *repeat.*

Ø reverse order

Redundancy. The word is *reverse.*

Ø with regards to

in regards to

Drop the *s.*

With **regard** to

In **regard** to

Ø thusly

Change to *thus.*

Thus, the center closed forever.

Ø toward

towards

Use either word to indicate direction.

Ø via

Latin for the English expression *by way of.* Use the English expression in running text or titles.

> Ø   The model was tested **via** computer simulations.
>
> The model was tested **by way of** computer simulations.

Ø -wise

Do not create constructions that end with *-wise.* Well-known words such as *clockwise* and *lengthwise* are acceptable, but fabrications such as *computerwise* and *wordwise* are not.

> Ø   **Manhole-wise,** every civil engineer knows that only circular covers can be trusted on city streets because any other shape can fall down into a shaft if it is incorrectly positioned.

Zero Lingo

For our last entry, let's look at the use of what are called "trend words" or "vogue words" such as *feedback, interface, bottom line, impact,* or *input.* Many of these are valuable, technical terms, but business magazines and the how-to-succeed industries popularize such terms, and they simply become silly buzz words.

In technical and scientific areas, our language is exploding by hundreds and possibly thousands of words a year. New and emerging fields create new and emerging vocabularies. A bioengineer cannot splice a gene without the vocabulary to define and explain the research. A new plastic or a new virus or a new household device calls for a new name. Anything we discover or create enlarges the language. Many of the popular new technical terms are abbreviations.

> The **CD** replaced the **LP.**

Some are acronyms (abbreviations made up of the first letter of each word) that we speak as words.

> The **laser*** improved inventory controls.
>
> A **LAN** is a local area network.

** **L**ight **a**mplification by **s**timulated **e**mission of **r**adiation*

Some are brand names.

Xerox and **Teflon** and **Kevlar** are trademarks.

These are acceptable words, and they are necessary elements in your writing as long as they meet industry standards, but they should not be used as buzzwords for human activities.

Ø   Look at Mr. Clean.

Ø   I need his input so we know where to go for lunch.

Ø   How will this impact Smith?

Ø   He couldn't access the client to finalize the deal.

Ø   Pencil me in and calendar Robin if you have a window.

Exercises Chapter 6, Section E

Correct the errors in the following sentences.

---

1. The 500 ships that were sunk off the east coast of the United States between 1940 and 1945 are now leaking and causing alot of pollution.

---

2. No other software menu features, as such, can challenge the publications design utility of the Adobe applications.

---

3. Close up you will see what is called "pixelation," which is the visible distortion caused by low pixel resolution.

---

4. Most automobiles, e.g., Fords and Hondas, are designed to be fuel efficient at 50 miles per hour rather than at 70 miles per hour.

---

5. Art Fry, a 3m engineer, developed a glue that sticks but doesn't and found it convenient to use on little slips of paper, e.g., notes.

---

6. The glue adheres one piece of paper to another, but its lack of adhesion strength makes it removable, i.e., temporary.

---

7. Outside of an air-conditioner, only a clothes dryer or a refrigerator consumes over 1000 kilowatt hours of domestic power a year.

_____

8. Zoomer introduced the first 200m zoom lens in 1946, but it was nowhere near as well designed as today's zoom lenses.

_____

9. The controversy over the relationship between meat eating and cancer risk is over with because studies show that vegetarians have roughly half the cancer risk of meat eaters.

_____

10. With regards to its energy reserves, the United States has only a 30-year supply of oil (in terms of present extraction technology).

_____

11. Thusly, the 200-year supply of coal will gain renewed attention in the near future.

_____

12. The 60-year supply of gas is nowhere near enough to handle the needs of the new century.

_____

13. Computer-wise, a new age has dawned.

_____

14. Will pesticide residuals impact the wildlife in the area?

_____

15. If the engineers can access Room 304 JT, we should have the new install completed by Thursday.

_____

CHAPTER 6   GOOFS, GAFFES, GLITCHES, AND GREMLINS

Zingers

You may not have noticed, but a number of the typical errors that writers make in their work are discussed several times in several chapters. This reinforcement may be helpful. To conclude the chapter, I repeat the list of the most notorious spelling errors.

Because rhyme-pair errors and look-alikes (see pp. 139–140) are so common, a grammar checker will flag some of them, but the program will not select the correct word. Once alerted, you must determine the correct word choice. Because the choice usually involves two options, half of your uses will be correct in any case. If you are prone to any of these misuses, you can also enter these words in a search feature on your computer.

Search-and-Execute Errors

to, too, two
their, there, they're
weather, whether
of, off
affect, effect
capital, capitol
who's, whose
principal, principle
accept, except
cite, site, sight
farther, further
its, it's
lead, led
loose, lose
choose, chose
your, you're
vary, very
than, then

Exercises Chapter 6, Section F

Correct the errors in the following sentences.

1. Although their are efforts to stabilize the famous leaning tower of Pisa, it is now tilted 17 feet, and this affect may topple the structure at some point.

2. All machines are based on six simple machines—or five if you take the wedge off the list.

3. Your at risk of absorbing aluminum from antacids and antiperspirants, but aluminum may or may not be a health hazard.

4. Ultrasound is a 3-D image technology that has no ill affects on the body.

5. The CAT scanner will show soft tissues at the desired cite.

6. X-ray technology can be a problem because of it's radiation levels.

7. *Outdoor enthusiasts have to chose between sunburn or safety.*

_____

8. *Skin damage can led to cancer.*

_____

9. *For many, the risk involved in ultraviolet radiation is hard to except.*

_____

10. *As we noted earlier, the shape of a manhole cover will never very for a vary simple reason; any shape other than a circle can fall into a vertical conduit.*

ESL Errors

By the year 2010, only 60% of the American workforce will be native speakers. A very large number of international students and immigrant students are enrolled in science, engineering, and engineering technology programs in both two- and four-year colleges, more than in most of the other campus programs. For this reason, there are a few areas of special concern for nonnative speakers, particularly students from China, Vietnam, and other Pacific Rim nations.

Campuses nationwide do their best to channel nonnative speakers through basic English grammar courses, but the challenge to the institutions as well as to the students can be overwhelming. The significance of basic language skills in English cannot, however, be overlooked or underestimated. There are many community college students who are unable to achieve their goals—not in school, but after graduation. Some students cannot pass the transfer English-proficiency exams to enter universities. Others who enter the workforce cannot secure the best jobs because corporations are particularly demanding when it comes to basic English skills. In the most regrettable cases, I find students in my classes who entered universities in which there were inadequate screening tests in English, and the students could not succeed in their programs. The message is simple enough: everything depends on your basic skills in math *and* language.

If you are a nonnative speaker, you can find some comfort in writing English documents in the scientific areas. The vocabulary is standardized and precise in every field from aeronautical engineering to zoology. Technical language is familiar to every student of EE or biomed or any other scientific or technical field, regardless of his or her social or cultural background. The math systems provide even more comfort, since the language of math is universal. In addition, the methods of organization and development in technical writing are predictable and repetitive. One problem remains: the English that holds it all together. If your command of English as a second language is reasonably strong, you should focus on four likely troublesome areas.

1) **Verbs** The familiar conjugations and tenses of English verbs do not exist in Pacific Rim languages. In fact, as linguists point out, there are tense concepts that exist in one culture and not another. In other words, the language for a verb form may not exist because the concept does not exist. You must memorize and practice your tenses and the conjugations so that the logic coincides with the logical expectations of the American reader. Ultimately, errors in this area become a logic barrier, and so the error creates confusion or misunderstanding. In any scientific consideration, you *must* choose the correct tense or the scientific observation may be

misunderstood. The past is the past. The present is the present. The future is the future, and so on.

2) **Articles** There are no articles in many of the Pacific Rim languages. The English language clutters every document with articles. To English speakers, the endless use of *a, an,* and *the* seems entirely necessary, even though the logical value of the articles is somewhat redundant, as you know. The word *lion* means a *lion* or *the lion*—but there is only one lion in either case. A native speaker will insist that *a lion* is not *the lion.* You must practice the use of articles because you cannot use any grammar check features for this task. Grammar checks examine what you *did* with the computer. Your tendency will be to *not* use the articles. The computer cannot easily identify what you did *not* do.

3) **Spelling** There are no alphabets in a number of the Pacific Rim pictographic languages. If you are used to reading characters, always use a spell checker. Be sure you establish a search file for technical terminology for your specialty. Memorize the basic rules of spelling (see pp. 133–144.)

4) **Pronunciation** Clear speaking is the best-kept secret of employment success for international and immigrant students. Grades matter, but speaking skills plus grades will guarantee the job. It can also be argued that pronunciation is the best-kept secret of writing success, and proper pronunciation will help get a writing job done, also! What you pronounce correctly you are more likely to spell correctly. If you pronounce the words correctly, you are more likely to use the words correctly.

Be alert to software advances in ESL. Ask your campus ESL program instructors if they know of any software that is dedicated to your particular ESL difficulties. Look for software that can grammar-check typical errors. You can be sure that, in time, there will be software applications dedicated to translations from Thai, Mandarin, Tagalog, and other languages. Watch for such programs in your community newsletters or newspapers and in magazines from your native country. Also look for handheld bilingual instruments. A number of Chinese students in my classes use electronic bilingual dictionaries that they carry to campus along with their calculators.

The problem areas are outlined next. Remember the Pareto principle and focus on important errors or frequent errors.

Verb Tenses

For many foreign speakers of any language, there is only one safe place to be when they speak—in the present tense. If you are always writing in the present tense, you need ESL coursework. From a scientific point of view, dwelling in the present tense may imply some technical precision in the content of instructions, but writing tends to use the past tense as a starting position. You need to be able to discuss the past and the future, and you must properly identify a process that is ongoing (progresssive).

The cell **collapsed.**

The bridge **will hold** the load.

The wires **are stabilizing** the tower.

Begin by clearly pronouncing the *ed* on the end of most past-tense verb, and the word *will* before the future-tense verbs. Talk with your hands and learn to point to the left for a past tense and to the right for a future tense. You are translating in your mind, and your finger will be faster than your speech selection. Follow your finger. Think of -*ed* on the left and the word *will* on the right. The past is on the left; the future is on the right. Oral practice will help you remember to write the endings you omit or are not hearing.

Also, when a subject is third-person singular, the verb takes an -*s* or -*es* ending in the present tense.

The **patient eats** vegetables as part of the therapy.

It is common to hear the *s* omitted in nonstandard English or in ESL speech patterns, and the error is often transferred to writing. Speak the *s* so that you remember to write the *s*.

Articles

The rules of article use are easy to explain, but they are difficult to practice if your native language does not use articles. Essentially, articles signal that the next word is a noun or object. The reference can be general or specific: *a* textbook or *the* textbook. The general or generic articles—*a* and *an*—cause the problems.

For General Objects

- you will use an article for the majority of nouns. Use *a* and *an* for generic objects that are singular *but* whose plural form ends in *s* or *es*. In other words, if the noun can take a plural ending (*s* or *es*) use *a* or *an* with the singular form. (There are only a few plurals, such as *oxen* or *mice,* that do not use the *s* endings.)

a pineapple   the pineapples

Use *a* before words beginning in a consonant

    a bear   the bears

and *an* before words beginning with a vowel sound.

    an eagle   the eagles

- For the plurals, always use *the,* as in the preceding examples. The plural form is quite simple.

For Specific Objects

- Use the article *the* to refer to specific nouns as opposed to generic items.

    I will have **an** apple.   (any apple, a generic reference to apples)

    I will have **the** orange.   (a specific orange)

For Precise Objects

- You can shift to another group of words called *determiners* if you want to be precise. (They are adjectives, not articles.)

    I will have **this** pear.

    I will have **that** peach.

For Mass Nouns

- The plurals of many English nouns are not constructed with *s* or *es* endings. "Mass nouns" can be quite confusing, and at least a hundred or more of them—from *research* and *information* to *soap, salt, rice,* and *water*—are words we use frequently. The list is far greater in science and technology, where you will see such mass nouns as *silver, copper, petroleum, plastic, equipment, cement, machinery,* and *gasoline.*

The error occurs in mistaking the noun as singular.

Ø   He conducted **a** research.

Do not use *a* or *an* with mass nouns. Use the article *the* or no article at all. The preceding sentence can be correctly stated in several ways.

    He conducted **the** research.

    He conducted research.

An additional problem here is that some mass nouns can also be plural and have *s* or *es* endings—*plastics, soaps, salts*—when seen as a count of how many: "How many soaps do you have in your house?" You cannot use the *a* or *an* with these words. Of course, the mass noun is similar to a regular plural in a way because both mass nouns and plurals refer to quantity. Mass nouns, however, refer only to quantity in volume (how much), whereas plurals state quantity by number (how many) (see also pp. 195–196).

Nouns

When I travel to a Spanish-speaking country and I am speaking Spanish, I am perfectly happy to recall a Spanish word without deciding whether it should be singular or plural. If you are establishing residency in the United States or if English is important in your career plans, the issue is more serious. If you frequently omit the plural forms, you will add more damage to any verb tense problems when you write because you will add what is called a "number" problem. A singular subject takes a singular verb, and a plural subject takes a plural verb, but if you do not add *s* to your plurals, you are not likely to recognize the plural, and then you will add the wrong verb.

Many language systems do not have (or the speakers cannot pronounce) the buzzing sound *zzzzz*. Practice it. You also need to practice imitating the sound of air leaking from a tire—the hissing sound *sssss*. If you speak these sounds, you will learn to write them. If you write them, you will eliminate two large sources of error from your work: number agreement errors and faulty verbs. Practice these sounds: Dog*zzz* are great fun. It work*sss*.

The machine twist**s** the caps.	(The *s* sound is at the end of the verb.)
He run**s** fast.	(The *z* sound is at the end of the verb.)
Do you want cornflake**s** or cupcake**s**?	(The *s* sound is at the end of the nouns.)
Where are the chair**s**?	(The *z* sound is at the end of the noun.)

Among Americans, the misuse of a verb tense or a conjugation can signal a negative social meaning. No one will apply that negative perception to a nonnative speaker, so you can be fearless in your efforts. Nonetheless, it is important to control the agreement problem by mastering verb agreement, plural nouns, and the correct pronunciation of verbs and nouns.

Singular Plurals

Electronics, mathematics, statistics, and *physics* are among a small group of plurals that are treated as singular forms. Treat them as you would treat the words *news* and *athletics*.

The good news **is** that electronics **is** a growing field.

In other words, do not refer to the career field of *electronics* as *electronic*.

Rhyme Pairs

Pairs. Pears. A native speaker will not confuse these two words. An ESL speaker has a much larger group of rhyme pairs (and pairs that nearly rhyme or otherwise sound similar) to worry about than the two lists I identified on p. 139–140. The rhyme problem is amplified for nonnative speakers, because of their limited exposure to the language or because of hasty reference to a dictionary. Unlikely errors, clearly caused by rhyming sounds, show up in ESL writing.

been	being	
buy	by	
he's	his	
due	do	
know	no	
seat	set	sit
were	we're	where
well	we'll	will

If you begin to maintain a list of these errors as they occur in your writing, you can memorize the correct usage and program a search for the words on your word processor.

Exercises Chapter 6, Section G

1. Spelling: Correct the spellings of the following words.

aquainted	mispell	preceed
beleive	morgage	prefference
comitment	nintieth	primative
disipation	noticable	procede
ellimination	obsticle	recieve
fullfil	occassional	referal
guidence	ocurrence	relevent
incidentaly	organazation	resistence
inovation	oscilate	seperate
labratory	paralell	unecesary

2. Verbs: Supply a verb in the tense requested.

We _____ the experiment.   (past)

The results _____ that the prototype is functional.   (present)

The funds for the project _____ for the rest of the year.   (future)

I _____ the procedure repeatedly.   (past)

Each demonstration _____ that the theory is correct. (present)

The practical applications _____ in the near future.   (future)

3. Articles: Fill in the blanks using the article a or an. In sentences that cannot be constructed with a or an, substitute the.

It was _____ telescope.

I dropped _____ evaporation dish on the floor.

_____ iron was oxidized.

There was _____ sample on the slide.

He needed _____ equipment.

_____ assortment of stones was in the bag.

4. Singular plurals: Develop a sentence that properly uses each of the following words.

electronic	statistic	physics	mathematics
electronics	statistics	news	

5. Plurals: Speak the sounds in the following two groups of plural words. Without the s and z sounds, the forms are singular.

Two instrument*s*	Two sample*s*
Three test*s*	Three viruse*s*
Four flight*s*	Four beaker*s*
Five result*s*	Five test tube*s*
Six assistant*s*	Six CRT*s*

Num- bering

The handling of numbers is complicated by the fact that they are commonly presented both in arithmetic characters as well as in written form. The usual dilemma is choosing between one and the other. A number of circumstances determine which convention is the more appropriate.

- Is the text technical or general?

- Are the numbers being used to quantify or measure technical or general items?

- How big or small are the numbers?

Two sets of guidelines govern the use of numbers. If the document you are developing is general or nontechnical, numbers appear with less frequency, and they are often rendered in words. In technical and scientific work, figures (numerals) are the preferred medium. A cover letter for a technical proposal might make use of spelled out numbers, and yet the proposal will more likely use numerals.

Sentence Logic
Sentence Symbols
Sentence Conventions
Word Rules
NUMBER RULES

General or Nonscientific Applications

- In everyday writing chores that are basically nontechnical, numbers that can be written in one or two words are spelled out. Notice that any number over twenty is hyphenated except for units of ten: *thirty, forty,* and so on.

  The first electric locomotives were demonstrated in Berlin in 1879. The first London subway trains used a pair of electric motors that generated **twenty-five** horsepower apiece with a **fifty** horsepower aggregate.

  Homo erectus, our ancestor, emerged over **one million** years ago.

- Numerals are used for a variety of applications in both general as well as technical documents. Use numerals for

Fractions	1/2
Decimals	0.009
Percentages	39% were males
Statistics	a 16-hand horse averages 1200 pounds
Time	3:15
Money	$4929.31
Dates	February 1, 2002
Addresses	3709 Sixth St.

Specific details for these uses will be explained in the following rules.

- In general, numerals are preferred if a *measurement* is involved.

  The wall was 21 **feet** long.

  The unit had a capacity of 30 **cubic feet.**

- Percentages are most conveniently rendered as figures with a percent symbol.

    30%

- In general writing situations, quantities, common measures, and fractions can *also* be written out, unless they are awkward.

    In troy weight, **twenty-four grains** equals one pennyweight.

    It was once thought that human reflexes could not handle vehicle speeds in excess of **twenty-five miles per hour.**

    A kilometer is approximately **three-fifths of a mile.**

    Commercial applications for roof construction are often composed of **six-by-twelve laminated beams.**

Any awkwardness is usually quite obvious.

    Ø   A common brick is seven and one-half inches long by three and one-half inches wide.

This sentence calls for numerals. In this instance, expressions such as "seven and one-half" are very awkward, so the fraction must be numerical.

    A common brick is 7-1/2" by 3-1/2."

- Larger numbers are usually expressed in numerals.

    Our moon is 240,000 miles from the earth.

- Large numbers can be expressed as a combination of numerical references *and* a large unit of measure expressed in words or, possibly, an abbreviation.

    Dinosaurs disappeared abruptly **65 million** years ago.

    Alpha Centauri, the nearest star, is **24 million million** miles from our sun.

- Do not begin a sentence with a numeral. Without a capitalized word, the signal for a new sentence is missing.

    Ø   **1600** or more asteroids compose the largest objects of the asteroid belt.

    Over 1600 asteroids compose the largest objects of the asteroid belt.

Readers depend on capitalization as a cue for a new sentence. If the number is conveniently expressed as a word or two, then you might write it out. If the number is best left as a figure, recompose the sentence to move the number.

Ø   **29** gallons escaped.

Twenty-nine gallons escaped.

The ship lost 29 gallons.

This rule holds true even in highly technical documentation. In fact, the more numbers are used, the less apparent the location of a new sentence becomes, so you must not start sentences with a numeral.

- Plurals of written numbers are no different from any other plurals. They are formed with the same *s* or *es* endings.

    sevens   eighties

- Plurals of numbers in their numerical form are also made by adding an *s* or, if preferred, *'s*.

    2 × 4s   30's

- Page numbers are *always* figures regardless of where or why they appear. In numbering a page and in mentioning a page, always use the number.

    page 31

- Decimals are *always* stated in numerals.

    31.029

- Fractions that are part of a number must appear as numerals. To say *30-1/2* as *30 and one-half* would be very awkward.

- Fractions are spelled out if they stand independently.

    One-third of the cost went for overhead.

- If two numbers are adjacent to each other, spell one of them out if it will help clarify the text.

    Ø   The convoy was composed of 6 231's.

    The convoy was composed of **six** 231's.

- Otherwise, if several numbers applicable to the same category appear in the same *sentence* they should be expressed in the same manner.

    There were **eighty** containers, but **thirty** were broken.

- If several numbers applicable to the same category appear in the same *paragraph,* they also should be expressed in the same manner.

A bat can be swung at 75 mph. A ball can be pitched at 95 mph. Should the ball directly meet the bat, the impact will deliver 8000 lbs of force. The subsequent trajectory will exceed 400 ft approximately 1% of the time. The other 99% is why many physics majors enjoy football.

Technical Applications

As the last two suggestions indicate, an increased incidence of numbers will usually call for numerals. *In technical applications, numbers are usually expressed in numerals.* This is particularly true if the numbers occur frequently. Technical material is simply easier to read in numerical format. After reading all the rules for the general use of numbers, you will probably be glad most of your material calls for numerals, but there are rules for the use of these as well. Rigorous use of numerals calls for special attention to a few particular rules that will be helpful.

- In technical documents, *quantities* are expressed numerically.

  8 amps

  60,000 cubic meters per hour

  25,000 megahertz

  The temperature in the mesosphere drops to −70°C.

  Above 500 kilometers is the layer known as the exosphere.

  The Trifid Nebula in the constellation of Sagittarius is about 3000 light-years away.

- Numerals must be used with any unit of measure that is *abbreviated* or expressed as a *symbol.*

  The human ear hears sound between 20 **Hz** (cps) and 20,000 **Hz.**

  The speed of sound in air is about 740 **mph.**

  High-quality printing calls for 1000 **dpi** capability.

- *Percentages* are stated in numerals with the % symbol.

  In relation to the number of women over age 65, the number of American men over age 65 *declined* by **45%** between 1900 and 1980.

  Commercial grades of clear glass windows transmit approximately **85%** of the available light.

- *Decimal fractions* are numerical.

  In the year 1900, New York City had a population of **3.6 million** people.

- American convention uses commas in large numbers. European convention often calls for periods to serve the same function. Scientific notation may delete *both* and use a space to divide large numbers.

30,126,132	American
12.131.882	European
3 419 373	International

The American convention is the most practical application for most scientific applications for domestic audiences.

These large-number symbols are positioned from the right, but the period may also function as a decimal point, in which case it is positioned from the left.

0.011

Also, be alert to the fact that European scientific documents may be using a comma as a decimal point.

- Parts numbers and other identification numbers are identified by numerals.

  95F3948WM shielded cable for **EIA/RS-232** applications

- The former practice of repeating a written number with the figure number is no longer in use.

  Ø There were five (5) oscilloscopes on the shelf.

Other Number Conventions

Dates

- Render dates by either the American convention

  On September 14, 1939, American aeronautical engineer Igor Sikorsky was at the controls for the world's first helicopter flight.

  or the European method, which is also the military convention.

  The first Soviet nuclear bomb was exploded on 22 September 1949.

- Do not write *th* after any reference to the day of the month. You speak the *th* but do not write it. April twelfth is written *April 12*. (See the last rule in the following "Geographic Matters" section for the optional use of *th* in correspondence.)

- You can spell out decades *or* use numerals.

  The surfin' sixties.

  The rockin' '70s.

  (Do not add a second apostrophe to an abbreviated numeral. It is an optional use, and it will create the awkward-appearing *'70's.*)

- Plural numbers are formed with *s* and *no apostrophe.*

  The cyclotron was developed in the 1930s.

- References to centuries are spelled out.

  The age of discovery began in the sixteenth century.

Geographic Matters

- Bearings are designated in numerals.

  39°N

- Highways are designated in numerals.

  Route 66

- Streets are most frequently spelled out.

  409 Thirty-Fifth Street

- In correspondence, the ordinal *th* is popular for addresses.

  5704 39th Avenue

Money

The guidelines for currency are similar to the preceding ones except that it is probably easier to identify awkward numbers in financial commentary. *Most tasks will call for numerals.*

- If you use numerals, use the $ symbol.

  $40,000

- If you spell out the numbers, spell out the symbols also.

  one million dollars

- For small dollar amounts, add two zeros. If you have five bucks, you have $5.00.

- For large numbers use a combination of numerals and the words *million, billion,* and so on. Use the currency symbol in this case.

> Once in production, each aircraft costs **$6 million.**

Written Numbers

Even though numerals are favored in technical documents, a few conventions are usually handled with words.

- Numbers from one thousand to ten thousand are often expressed in words if the units are stated in thousands.

> A hurricane struck Galveston, Texas, on September 8, 1900, and the storm killed as many as **eight thousand** people. This stands as the greatest natural disaster in American history, one that would not be likely to recur in our age of climatology and storm-watch technology.

- Ranks (ordinal numbers) from the first to the ninety-ninth are spelled out.

> Twenty-three planetary moons were discovered before the aid of photography. The **twenty-third** moon to be identified was Amalthea, which was also the **fifth** moon of Jupiter to be discovered.

CHAPTER 7 NUMBERS, MEASURES, ABBREVIATIONS, AND SYMBOLS

Exercises Chapter 7, Section A

Correct the errors in the following sentences. Assume that these sentences come from a nontechnical document.

1. The first commercial computer in the U.S. was UNIVAC 1, and it contained 5000 vacuum tubes.

2. During the 20th Century, Americans added 28 years to their longevity.

3. The 28 years was a greater gain in lifespan than humans had experienced in the preceding 2 thousand years.

4. Over sixty percent of all garbage was dumped in landfills in 1960.

5. Hummingbirds can fly 70 mph.

6. 15 years ago, Thinking Machines introduced a computer that had a speed of 1 billion operations per second.

Assume that the following sentences come from technical documents. Make the appropriate corrections.

7. On an average day the heart rate is seventy-two beats a minute, which means that the heart will beat one hundred thousand times in twenty-four hours.

8. A stack of recycled newspapers four feet high will save a forty-foot tree.

9. American garbage is twenty-three percent paper, but that percentage (equal to nineteen hundred pounds a year per citizen) is declining as a result of effective recycling.

10. Scientists believe a quasar emits as much light as a thousand galaxies and produces as much energy as one hundred trillion stars.

11. The precious-stone measure called the carat equals two hundred milligrams or one hundred points.

12. When we consider the density of water, it is amazing that the cosmopolitan sailfish (Istiophorus platypterus) can travel at sixty mph.

13. As early as 1974, Panasonic demonstrated an HDTV with over eleven-hundred scanning lines.

Measures and Abbreviations

Abbreviations are the stock-in-trade of science and industry. They are time- and space-savers, but they do depend on the educated awareness of a reader who learns the abbreviations that are standard protocol in any given discipline, whether it is civil engineering or astrophysics. There are also, of course, a wide variety of abbreviations that are either conventions or conveniences for general writing practices.

The proper use of abbreviations and their correct spellings are very important for the success of a technical or scientific document. Abbreviations must be used as precisely as the signs of mathematics or the symbol codes of an engineering field. Misspelling a word is regrettable; misusing an abbreviation could be a monumental error. A misspelled word is not likely to lead to a systems failure, but incorrect abbreviations could guarantee it. Errors in abbreviations or symbols could result in the device that malfunctions, the formula that fails, or the program language that defaults. Miscalculations have cost many lives in the pursuit of engineering achievements, among early jet test pilots, for example. Miscalculations have flattened more than one lab or test site, including the famous Alfred Nobel labs, where he sought the discovery of what was later called *dynamite*. The astronauts who have been lost were victims of engineering failures.

There are two general uses for abbreviations:

1) Scientific documents make extensive use of abbreviation systems as *shortcuts* to articulating extremely complex considerations.

2) A variety of practical abbreviations help us *codify* detailed printed material—such as indexes, inventories, or other lists—and offer practical conventions for such workaday tasks as writing mailing addresses.

There are three elements to consider in correctly applying the use of abbreviations.

- The accepted standard spelling of the abbreviation.

- When to capitalize, which is usually a matter of convention or standard practice.

- When to add periods, which may also be part of the accepted standard. (This seems to be a practice that is losing ground, as I noted elsewhere.)

It is important for you to identify a reliable source for the abbreviations used in your area of specialization. Professional associations in engineering and scientific fields have standardized many systems of abbreviations. You might also be able to locate practical software applications that can spell-check your use of abbreviations in your field. The following general guidelines may be of interest.

Standard International Measures

The International System of Units (SI) is the current world standard of measures used in the sciences and technologies. It is based on the metric system. Seven *basic units* form the foundation of the SI system (the abbreviation is from the French *Système International*).

Quantity	Unit	Abbreviation
length	meter/metre	m
mass	kilogram	kg
time	second	s
electric current	ampere	A
thermodynamic temperature	kelvin	K
amount of substance	mole	mol
luminance intensity	candela	cd

Fractions and multiples of the base units are created by adding prefixes from the following table to the base units. Thus, such terms as *nanosecond* and *picosecond* are derived by combining a prefix and a base unit. The prefixes are increasing and decreasing powers of ten.

Prefixes

Factor	Prefix	Symbol	Factor	Prefix	Symbol
10^{24}	yotta	Y	10^{-1}	deci	d
10^{21}	zetta	Z	10^{-2}	centi	c
10^{18}	exa	X	10^{-3}	milli	m
10^{15}	peta	P	10^{-6}	micro	μ
10^{12}	tera	T	10^{-9}	nano	n
10^{9}	giga	G	10^{-12}	pico	p
10^{6}	mega	M	10^{-15}	femto	f
10^{3}	kilo	k	10^{-18}	atto	a
10^{2}	hecto	h	10^{-21}	zepto	z
10^{1}	deka (or deca)	da	10^{-24}	yacto	y

As radio telescopes and electron microscopes enlarge our universe we need ever larger and ever smaller units of measure, such as the *light-year,* the *parsec* and the *femtometer* (also called a *fermic.*) Numbers between 0.1 and 1000 are acceptable in modern documentation but you will usually not see large numbers except as a curiosity. Instead, the system of powers is used to abbreviate the logic of the unit. The practical logic is apparent at the modest level of 1000 grams, which equals 1 kilogram; however, the utility is more obvious when dealing with the "Megs" on your computer or the theoretical weight of a planet—or even a star.

Powers are traditionally expressed with a superscript to the upper right of the basic unit symbol, and a variety of these abbreviations are commonplace in international standards. The following are some standard *derived units*.

Quantity	Unit	Symbol
area	square meter	m^2
volume	cubic meter	m^3
velocity	meter per second	m/s *or* $m{\cdot}s^{-1}$
acceleration	meter per second squared	m/s^2 *or* $m{\cdot}s^{-2}$
density	kilogram per cubic meter	kg/m^3 *or* $kg{\cdot}m^{-3}$
luminance	candela per square meter	cd/m^2 *or* $cd{\cdot}m^{-2}$
heat capacity	joule per kelvin	J/K *or* $J{\cdot}K^{-1}$
dynamic viscosity	pascal second	$Pa{\cdot}s$

The names of the units are usually not capitalized (although a number of them are derived from a name, usually that of a discoverer such as André Ampère, Heinrich Hertz, James Prescott Joule, and Alessandro Volta among others you might recognize).

- Note that there is *no* plural form.
- Note that the abbreviations are used *without* periods.

A number of SI equivalents have their own name. Note that these are capitalized symbols.

Quantity	Unit	Symbol	Equivalent
frequency	hertz	Hz	cycles per second
force	newton	N	kilogram-meters per second squared
pressure	pascal	Pa	newtons per square meter
energy	joule	J	newton-meter
power	watt	W	joules per second
quantity of electricity	coulomb	C	ampere-second
electric potential	volt	V	watts per ampere
capacitance	farad	F	coulombs per volt
electrical resistance	ohm	Ω	volts per ampere
electrical conductance	siemens	S	amperes per volt
magnetic flux	weber	Wb	volt-second
inductance	henry (pl. henries)	H	webers per ampere
absorbed dose	gray	Gy	joules per kilogram
activity (of radionuclides)	becquerel	Bq	disintegrations per second

You may not be familiar with many of these terms, but they are important parts of the vocabulary of engineers and scientists.

> One of the fundamental problems in the development of fusion reactors is that fuel ignition takes almost five hundred terawatts of electricity generating two million joules of radiation at a temperature of three million degrees for four nanoseconds.

English Measure

You can see that the scientific community favors international, metric, and discipline-specific systems of measure and abbreviation. The conspicuous exception to this is the construction industry and related disciplines, such as architecture, drafting, surveying, and associated fields in civil, mechanical, and construction engineering. Traditional measures are very much in place in these environments—at least in the United States.

Length		Area		Volume	
in. *or* "	inch	sq. in.	square inch	cu. in.	cubic inch
ft. *or* '	foot	sq. ft.	square foot	cu. ft.	cubic foot
yd.	yard	sq. yd.	square yard	cu. yd.	cubic yard
rd.	rod	sq. rd.	square rod		
mi.	mile	sq. mi.	square mile		

Traditional measures of weight and capacity also remain popular, including the very nonmetric pound and quart. Use a period with abbreviations for popular English measures if desired, but the punctuation is usually omitted in scientific applications.

Weight		Dry Measure		Liquid Measure	
gr.	grain	pt.	pint	min.	minim
s.	scruple	qt.	quart	fl. dr.	fluid dram
dr.	dram	pk.	peck	fl. oz.	fluid ounce
dwt.	pennyweight	bu.	bushel	gi.	gill
oz.	ounce			pt.	pint
lb. *or* #	pound			qt.	quart
cwt.	hundredweight			gal.	gallon
tn.	ton			bbl.	barrel

Again note that there are no plural forms for abbreviations of units of measure. If the beaker contains ten ounces, it contains 10 oz. If you have a bushel and a peck, you have 5 pk.

There are a great many abbreviations that characterize specific disciplines. For historical and practical reasons, these terms, particularly the units of measure, retain their popularity although they are not SI standards. The following examples are used in a number of technical and scientific fields.

ac	alternating current
AF	audio frequency
AM	amplitude modulation
bar	unit of atmospheric pressure
BHP	brake horsepower
BP	boiling point
Btu	British thermal unit
cal	calorie
Cal	kilocalorie
cos	cosine
CP	candlepower
ctn	cotangent
dB	decibel
emf or EMF	electromotive force
hp	horsepower
kW	kilowatt
Mc	megacycle
MP	melting point
MPH	miles per hour
NS	not significant
pH	alkalinity/acidity
rad	radian
rev	revolution
RF	radio frequency
rpm	revolutions per minute
SD	standard deviation
SE	standard error
sr	steradian
tan	tangent

Industry Abbreviations

Each industrial sector develops codes that evolve out of the special mathematical or engineering features that characterize an industry. Even the spray paint used to designate utilities on the street in front of your house is color coded by civil engineers:

blue	→	**water**
yellow	→	**gas**
green	→	**sewer**

| orange | → | **telephone** |
| **red** | → | **electric** |

The abbreviations often are directly related to the symbols used in engineering drawings. Designers and engineers who specialize in computer architecture use the following abbreviations. They are directly related to the symbols on p. 251.

AND	**Boolean AND**
ANDCC	**Boolean AND, set icc**
ANDN	**Boolean NAND**
ANDNCC	**Boolean NASD, set icc**
OR	**Boolean OR**
ORCC	**Boolean OR, set icc**
ORN	**Boolean NOR**
ORNCC	**Boolean NOR, set icc**
XOR	**Boolean exclusive OR**
XORCC	**Boolean exclusive OR, set icc**
XNOR	**Boolean exclusive NOR**
XNORCC	**Boolean exclusive NOR, set icc**

You will, of course, become familiar with the specialized abbreviations in your field.

Chemical Elements

The chemical symbols for most of the elements are abbreviations consisting of one or two letters, based on their Latin names and *not* their common names in English. Some of the common names also have changed over the years, and "wolfram" became "tungsten" and "quicksilver" became "mercury." Three elements recently obtained by nuclear reactions—Unh, Unp, Ung—are abbreviated with three letters. Note that symbols for the elements are *not* followed by a period. A great many engineering and scientific fields depend on the abbreviations of the elements.

The Elements

actinium	Ac	helium	He	radium	Ra
aluminum	Al	holmium	Ho	radon	Rn
americium	Am	hydrogen	H	rhenium	Re
antimony	Sb	indium	In	rhodium	Rd
argon	Ar	iodine	I	rubidium	Rb
arsenic	As	iridium	Ir	ruthenium	Ru
astatine	At	iron	Fe	samarium	Sm
barium	Ba	krypton	Kr	scandium	Sc
berkelium	Bk	lanthanum	La	selenium	Se
beryllium	Be	lawrencium	Lr	silicon	Si
bismuth	Bi	lead	Pb	silver	Ag
boron	B	lithium	Li	sodium	Na
bromine	Br	lutetium	Lu	strontium	Sr
cadmium	Cd	magnesium	Mg	sulfur	S
calcium	Ca	manganese	Mn	tantalum	Ta
californium	Cf	mendelevium	Md	technetium	Tc
carbon	C	mercury	Hg	tellurium	Te
cerium	Ce	molybdenum	Mo	terbium	Tb
cesium	Cs	neodymium	Nd	thallium	Tl
chlorine	Cl	neon	Ne	thorium	Th
chromium	Cr	neptunium	Np	thulium	Tm
cobalt	Co	nickel	Ni	tin	Sn
copper	Cu	niobium	Nb	titanium	Ti
curium	Cm	nitrogen	N	tungsten	W
dysprosium	Dy	nobelium	No	unnilhexium	Unh
einsteinium	Es	osmium	Os	unnilpentium	Unp
erbium	Er	oxygen	O	unnilquandium	Unq
europium	Eu	palladium	Pd	uranium	U
fermium	Fm	phosphorus	P	vanadium	V
fluorine	F	platinum	Pt	xenon	Xe
francium	Fr	plutonium	Pu	ytterbium	Yb
gadolinium	Gd	polonium	Po	yttrium	Y
gallium	Ga	potassium	K	zinc	Zn
germanium	Ge	praseodymium	Pr	zirconium	Zr
gold	Au	promethium	Pm		
hafnium	Hf	protactinium	Pa		

Exercises Chapter 7, Section B

Correct the errors in the following sentences.

1. Studies reveal that 100,000 wind generators of the 1000 Kw variety could supply 15 percent of the nation's electricity.

2. CD's revolve at a much higher r.p.m. than conventional analog devices.

3. One form of steel is an alloy that includes chromium (   ) and vanadium (   ).

4. The mixture requires 5 oz's of styrene.

5. The speakers used eight ohm per channel.

6. Today's airports are often huge in area. (10 square mi. or more

7. Does anybody listen to am radio anymore?

8. The kilogram (Kg) is a unit of mass in physics.

_____

9. The square meter (M^2) is a standard unit of commerce in Europe.

_____

10. The unit of measure for thermodynamic temperatures is the Kelvin.

_____

11. It took 3 pts of resin additive to prepare the 5-gal drum of epoxy base.

_____

12. The area contained 3000 sq/yds of radioactive debris.

_____

13. HVAC calculations often involve the British thermal unit (BTU).

_____

14. What is the PH of water?

_____

15. As a result of French research, Radium (   ) became known throughout the world.

_____

16. Units per steel drum = 55 ga.

_____

General Abbreviations

There are hundreds of abbreviations that are frequently used for situations of a less technical nature. You will need to identify a source for abbreviation standards appropriate to your field, but you should also be familiar with the group of abbreviations that you are likely to encounter on a daily basis in your work. They are everywhere around us. You could hardly address an envelope without them.

Titles

- Degrees are often identified as a type of academic title after a name. They are widely used in medicine.

D.D.S.	Doctor of Dental Surgery
D.O.	Doctor of Osteopathy
M.D.	Doctor of Medicine

Such abbreviations are less widely used among scientists and engineers, who are more often addressed simply as Dr. Phillip Levine, for example. It is assumed nowadays that professors are doctors and that most of these nonmedical doctors are *Ph.D.'s,* a catchall phrase that rarely means a doctor of *philosophy,* which is what the abbreviation indicates. In general, bachelor degrees and master degrees are not commonly identified after a name in the United States.

States

- There are two sets of standard state abbreviations in the following table. The original abbreviations are on the left in each column, and they remain popular. Use the two-letter postal code for envelopes and mailing labels.

Ala.	AL	Ky.	KY	Ohio	OH
Alaska	AK	La.	LA	Okla.	OK
Amer. Samoa	AS	Maine	ME	Ore.	OR
Ariz.	AZ	Md.	MD	Pa., Penn.	PA
Ark.	AR	Mass.	MA	P.R.	PR
Calif.	CA	Mich.	MI	R.I.	RI
Colo.	CO	Minn.	MN	S.C.	SC
Conn.	CT	Miss.	MS	S.Dak.	SD
Del.	DE	Mo.	MO	Tenn.	TN
D.C.	DC	Mont.	MT	Tex.	TX
Fla.	FL	Nebr.	NE	Utah	UT
Ga.	GA	Nev.	NV	Vt.	VT
Guam	GU	N.H.	NH	Va.	VA
Hawaii	HI	N.J.	NJ	V.I.	VI
Idaho	ID	N.Mex.	NM	Wash.	WA
Ill.	IL	N.Y.	NY	W.Va.	WV
Ind.	IN	N.C.	NC	Wis.	WI
Iowa	IA	N.Dak.	ND	Wyo.	WY
Kans.	KS				

Many of the old abbreviations are more identifiable because they were often based on the first syllable of the state name, a quite unmistakable group of syllables. The other list, the result of a bureaucratic decision-making process in Washington D. C., gives us the somewhat confusing two-letter system of

ME, MD, MA, MI, MN, MS, MO, MT

The two-letter system predates the zip codes and barcodes, which probably left the code antiquated in any case. Use the full name of the state in the running text of correspondence and other documents that mention states, or cities and states.

He lived in Pennsylvania.

They lived in Houston, Texas.

- The names of countries are often abbreviated

U.K.   U.S.

- Addresses are often abbreviated, but in running text an address should not contain abbreviations as a general rule, except for these four points of the compass: *NW, NE, SE,* and *SW.*

The campus is located on College Way North, but the lab is four miles away on Madronna Drive NW.

Businesses

- Company names are usually written in full in the body of a letter or similar document, but certain abbreviations are frequently used in any bibliography, index, or other source that is compiled as a list. These include

    Co.   Corp.   Inc.   Ltd.   S.A.

- Many associations, agencies, organizations, and a host of other groups come to be known by their abbreviated form, which often forms an acronym.

    AAAS   American Association of the Academy of Sciences

    NASA   National Aeronautics and Space Administration

Latitude and Longitude

- If cited as bearings or coordinates, these terms are abbreviated or omitted. In general commentary, they are *not* abbreviated.

    lat   41°   30′   15″   N

               41°   30′   15″   N

    long   80°   15′   30″   E

               80°   15′   30″   E

    The **latitude** of Portland, Oregon, should create a colder climate in the winter months, but the air currents from the tropical **latitudes** of the Pacific Ocean protect the area.

Months and Days

- Standard abbreviations are as follows. Avoid the shorter forms indicated; they are used in very limited applications.

Jan.	Feb.	Mar.	Apr.	May	June	July	Aug.	Sept.	Oct.	Nov.	Dec.
Ja	F	Mr	Ap	My	Je	Jl	Ag	S	O	N	D

		Sun.	Mon.	Tues.	Wed.	Thurs.	Fri.	Sat.
		Su	M	Tu	W	Th	F	Sa

Time

- The conventions for standard time are used worldwide, although you must remember that a clock can be perceived as units of twelve (a.m. and p.m.) or a unit of twenty-four, as in Europe, for example,

    NBC news will be on at 11:00 p.m.

    The BBC World Service news will be on at 18:00.

There are a number of standard abbreviations.

sec.	**second**	
min.	**minute**	
h. *or* **hr.**	**hour**	
d. *or* **day**	**day**	
mo.	**month**	
yr.	**year**	
a.m.	***ante meridiem***	**(before noon)**
p.m.	***post meridiem***	**(after noon)**

Abbreviations of Popular Expressions

- Avoid the following two popular Latin abbreviations that are often found in formal writing. The English equivalent is precise and assures you that every reader will get the message.

e.g.	(Latin *exempli gratia*)	for example
i.e.	(Latin *id est*)	that is

- Avoid the abbreviation *etc.,* avoid the Latin version *et cetera,* and avoid the expression *and so forth.* All three are commonly misspelled, and none of them function with much precision in scientific contexts any more than they would on shipping orders.

Protocols

- The common titles that are attached to surnames are always abbreviated, and most take a period.

    Mr.   Ms.   Miss   Mrs.

They are always used in correspondence, but they are seldom used in papers, essays, or similar documents that refer to engineers and other professionals by name.

Dear Mr. Smith:

I recently read your letter . . . .

The chemist Linus Pauling popularized the consumption of large doses of vitamin C.

Resumes

- Avoid abbreviations in a resume unless the alternatives are uncommon or unconventional. Abbreviations seem logical enough for a document that runs only a page or two, but excessive use of them leaves the resume dense, choppy in style, and clipped in content. Be certain the abbreviations that you use are standard professional versions that will be clear to intended readers. Finally, remember that the resume is supposed to build a detailed image of you. Proper nouns help build that image. If you abbreviate the nouns, you may diminish the desired image.

  Ø   Sysop

  Ø   AAS/HVAC

These abbreviations should be stated in words:

Systems Operator

Associate of Applied Sciences Degree in Heating, Ventilation, and Air-Conditioning Technology

Exercises Chapter 7, Section C

1. Provide the correct conventions for the following.

Robert Whitaker, B.A., developed the theory.

D.D.S. Harold Bell, was among the first to use ceramic fillings.

2. Provide the older abbreviations for the "M" group of two-letter abbreviations for the states.

ME	MN
MD	MS
MA	MO
MI	MT

3. Add the correct abbreviations to the following sentences.

He worked for the Internal Revenue Service (   ).

She worked for General Electric (   ).

They worked for the Bonneville Power Authority (   ).

4. Correct the following sentences:

_____

We won't meet until 3:00 P.M.

_____

Good afternoon, Phillips. Come in and sit down.

_____

Crowthers is an R&D VP, or at least that is what it says here on his resume.

_____

If there is no zip code on the envelopes, we get lots of mail in Anchorage that has AK incorrectly written on the envelope instead of AL.

Symbols

Symbols are shortcuts for use in mathematical systems, chemistry, physics, and a wide variety of engineering and technical applications. In the math and science areas, the symbols can be used in the running text, but they will frequently appear in indented *displays* instead (see Chapter 8). In engineering and technical areas, the symbols frequently appear in engineering drawings. Here again, the symbols are not extensively used in the running text. The math symbols that you are likely to need in engineering and engineering technology areas are included in the following list. Although these symbols can be rendered by hand, your computer software should provide you with options, particularly an equation editor.

Standard Signs and Symbols: Mathematics

Symbol	Meaning	Symbol	Meaning
+	Plus	″	Second
−	Minus	°	Degree
×	Multiplied by	√	Square root
÷	Divided by	∛	Cube root
=	Equal to	∜	Fourth root
±	Plus or minus	ⁿ√	nth root
≡	Identical with, congruent	Σ	Summation
≠	Not equal to	Π	Product sign
≈	Nearly equal to	π	Pi
<	Less than	∪	Union sign
>	Greater than	∩	Intersection sign
≮	Not less than	!	Factorial sign
≯	Not greater than	∅	Empty set/null set
≤	Less than or equal to	∈	Is an element of
≥	Greater than or equal to	e	Base or natural logarithms
⊂	Included in	e	Charge of the electron
⊥	Perpendicular to	Δ	Delta
∠	Angle	∝	Variation
∟	Right angle	h	Planck's constant
△	Triangle	ℏ	h/2π
⌢	Arc	k	Boltzmann's constant
⌓	Sector	∂	Partial differential
∴	Hence, therefore	∫	Integral
·	Multiplied by	∮	Contour integral
:	Ratio	∿	Cycle sine
′	Minute		

Each discipline has a wide assortment of symbols that conveniently structure the symbolic formulations characteristic of the field. Chemistry, for example, has a particular format for rendering compounds in terms of the composition of the elements. The entire periodic table of elements is integral to the system. I noted earlier that many scientific and engineering presentations use the abbreviations for the elements because chemistry is often a component of the other science and engineering fields. A number of disciplines, such as astronomy or microbiology, might use the following symbols for a more thorough elaboration of chemistry in terms of bonds and reactions. Together with the table of element symbols, the following signs can structure the logic of chemistry.

/	**Single bond**	╲	**Double bond**
│	**Single bond**	‖	**Double bond**
\	**Single bond**	⫽	**Double bond**
│	**Single bond**	↔	**Reaction goes and left right**
≡	**Triple bond**		
⇋	**Equilibrium reaction**		
↕	**Reaction goes both up and down**		

There is no practical, single guide to scientific and engineering symbologies. You will ultimately become quite familiar with the symbols that are appropriate to your field. As an example of the enormous variety of symbol systems, look at a very small sampling of the symbols utilized in electrical engineering drawings for a host of applications. First, there are the traditional symbols widely used in the electronics industry.

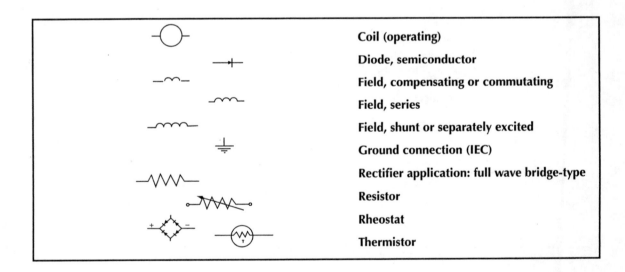

	Coil (operating)
	Diode, semiconductor
	Field, compensating or commutating
	Field, series
	Field, shunt or separately excited
	Ground connection (IEC)
	Rectifier application: full wave bridge-type
	Resistor
	Rheostat
	Thermistor

There are equally specialized logic symbols used in digital electronics for gates and other digital configurations.

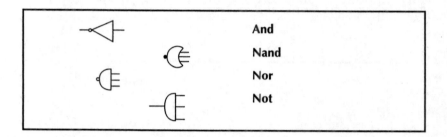

And

Nand

Nor

Not

On the higher voltage end of the industry, there are symbols for power applications.

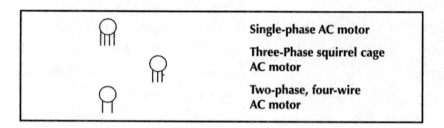

Single-phase AC motor

Three-Phase squirrel cage AC motor

Two-phase, four-wire AC motor

There is yet another group of symbols for commercial wiring applications. For example, here are a few symbols for receptacles.

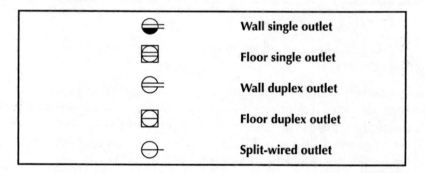

Wall single outlet

Floor single outlet

Wall duplex outlet

Floor duplex outlet

Split-wired outlet

This one engineering area illustrates that there is little or no carryover of symbols from one discipline to another except for the roles that might be played by mathematics and chemistry. In fact, as you can note in the four different groups of symbols from electrical engineering industries, applications of symbols are even unique *within* a specific engineering field: consumer electronics, digital electronics, power and instrumentation, commercial wiring, and so on.

Control panel codes and keyboard codes are excellent examples of the way in which technical processes are given a technical vocabulary that is, in turn, abbreviated or structured in symbols. Our knowledge of the abbreviations becomes *the* link that connects us to a great many sophisticated pieces of equipment. Here is the abbreviation key for the equation writer of the HP-485x Scientific Calculator.

Key	Operation
▲	Starts the numerator
▷ or ▼	Ends a subexpression
◄	Invokes the Equation Writer selection environment
← ()	Enters parentheses
SPC	Enters current separator
EVAL	Evaluates the equation and exits Equation Writer
ENTER	Returns the equation to the stack
ATTN	Exits Equation Writer without saving the equation
← GRAPH	Invokes scrolling mode
← EDIT	Returns the equation to the command line for editing
STO	Returns the equation to the stack as a graphics object
← { }	Turns "implicit parenthesis" mode off
→ " "	Returns the equation to the stack as a string

The code is, indeed, the *key* to understanding and operating a great deal of instrumentation in today's industries.

Because the symbols function as a language of sorts, they will appear in a wide variety of applications that vary from equations and calculations to chemical formulations to circuit diagrams to wiring blueprints. The necessary accuracy of the symbols is obvious enough, but proofreading the material should be an important focus of attention. Remember that proofreading includes the symbols and their proper utilization.

I always suggest to authors that a technical document needs *two* editorial readers. One reader reads the text for the language conventions you see throughout the *Writer's Handbook*. The other reader looks at the engineering details, particularly the accuracy of the math and the drawings and the symbols used in these engineering activities. One editorial skill is no less important than the other.

Exercises Chapter 7, Section D

1. *Develop a one-page symbology listing a series of frequently used symbols in your area of specialization. Identify each symbol. The symbology should contain ten to twenty symbols if possible. If you identify only ten, add a written definition for each symbol.*

2. *Create a screen scan of an important icon bar and explain each icon in a sentence or two. Illustrate each icon to the left of its explanation.*

3. *Interpret the operation keys on your math calculator, or focus on one group of them that is used to carry out a certain set of functions. Illustrate each key and place the explanation beside the illustration.*

Mathematics and the Written Text

Technical writing involves a specific type of project management that is unique to scientific and engineering communication. The central task is the management of the languages that are used to calculate and articulate the subject matter, which can range far and wide, from inorganic chemistry to digital electronics to construction engineering. The use of a written text in English is the communication medium that will hold the rest of the material in suspension—a well-organized suspension it is hoped. The rest of the material consists of radically divergent systems of logic and the language structures of those logic systems.

Each of the math systems is unique, from Boolean algebra to binary numbers, from real and complex analysis to numerical analysis, and newer areas of interest such as optimization and discrete algorithmic mathematics. The formula methodologies of disciplines such as chemistry are also unique, and a host of graphic features in chemistry also function somewhat as language. These many languages complement and support the entire body of work that is generated by engineers, scientists, and technicians.

The problem unique to this writing activity is the process of fusing the languages. An organic chemistry paper that uses chemical formulas, ball-and-stick conceptual illustrations, electron micrographs, higher mathematics, and English is operationally pentalingual (uses five languages). The challenge is to smoothly integrate the various materials, which is time consuming. One of the major style guides, *The Chicago Manual of Style,* reports that technical documents—the ones with all the calculations—are called *penalty copy* among printers and typesetters in the trade. Nobody wants to deal with the material, as any engineer or technician knows! In truth, however, computers have greatly eased the burden.

After examining the problem, I will explain ways to cut down on the punitive task of constructing symbolic systems on a page of text. First, there is a very real and mechanical reason why words do not blend well with formulas or calculations of any sort. Languages of the word-based variety evolved as spoken languages. These structures are sequential and, when written, are strung together straight across a line. Moreover, they usually have short or fast forms—called script—to speed the process along. In the science and engineering disciplines, authors tend to shift much of the heavy thought-processing into the math-based symbol systems that support other logic structures. The mathematical constructs that explain quantum mechanics, for example, do a job that English cannot do at the theoretical mathematical level.

Although chemists can say and write *water,* they cannot use the word with much success as a symbol, so chemistry, like math, structures its own language for the same word: H_2O. The symbol for H_2O is derived from the symbols for oxygen and hydrogen from the periodic table. The symbol can be drawn so that the word water can be visualized.

The method of presentation in the preceding figure is referred to as a *structural formula*. In other words, there is a symbolic language that is used in chemistry, and there is also an abbreviated form referred to as *chemistry notation*. The atomic structure of a compound is equally apparent in chemistry notation. Abbreviations like H_2O are a convenience and a trait all languages share; so tetra phosphorus decaoxide simply becomes P_4O_{10}. Many chemical names are far longer; chemists need the abbreviations.

The symbolic systems constructed for any given field of math or science or engineering are codes that best serve the needs of the discipline. Unfortunately, the interfacing of any two of these systems can be a point of confusion. One of the basic reasons for the problem is that English follows the simple trail of written words on a single line on a sheet of paper—which is then followed by another line. Printing equipment is designed for that dominant simplicity. Math systems, however, generally involve a double-line format because division is often involved in the calculations. At times, very complex concepts can be handled on one line, as in this calculation* for voltage drop.

$$-Rin(Ib) - Vbe - Re(2*Ie) + Vee = 0$$

However, you are more likely to see a stacked calculation, such as the following example from the American Society of Agricultural Engineers.

$$\frac{\Delta \rho}{L} = X_1 + X_2 \frac{\left(\rho_b / \rho_k\right)^2 Q}{\left[1 - \left(\rho_b / \rho_b\right)\right]^3} + X_3 \frac{\left(\rho_3 / \rho_k\right) Q^2}{\left[1 - \left(\rho_b / \rho_k\right)\right]^3}$$

This is a calculation for estimating the pressure drop in a bulk-processing procedure for shelled corn (ASAE Standard D272.2).

The solution to mixing a single-line language system and a multiline symbol system is to keep them separated. Since they do not occupy the same space in the same way, you must shift the page layout back and forth from one to the other. This is why you always see the math or chemistry or other symbolic systems set off on a page of text. First, a printer, either human or automatic, cannot structure a symmetry that both English and math can share. Second, and more important, it is better to keep the two separated. It is much easier to read a page that clearly indicates shifts in and out of the logic systems, even if it looks fragmented. The page becomes more readable if you do not combine the systems on a typed line. In chemistry, for example, you could write 1/[A]=Kt+1/[A]0, but it is probably clearer to omit the slash and display this law for a second-order reaction.

$$\frac{1}{[A]} = kt + \frac{1}{[A]_0}$$

Do you notice how much clearer it is if you let layout and white space do the work? The over-and-under format is visual in the second instance and symbolic in the first, and the topographical device of centering and surrounding the symbolic system with white space clearly delineates the law. Like so much of what you examine in this text, many of these practices are very well handled in the college textbooks you read, but when you try to write, you will not necessarily be able to bring those practices or procedures into play. Perhaps the layout logic is not apparent, or perhaps a textbook example is not evident without discussion.

*The use of boldface is occasionally used to highlight this discussion of math and chemistry. Do not boldface your math, chemistry, or physics displays.

There are practical guidelines for integrating math into a text. First, you have three options for presenting your work at the symbolic level. It is still perfectly acceptable to present the material manually, with a pen. Simply enter the calculations at the appropriate location on the typed page. Ink is superior to pencil but harder to correct. Leave extra space, since hand-lettering is always larger than type. Avoid blue ink. Use black.

The results can be seen in equations (3) and (6). The actual values are designated on figure 1 and as follows:

$$R_{L_1} = 544\Omega\ (495\Omega + 49\Omega)$$
$$R_{L_2} = 542\Omega\ (493\Omega + 49\Omega)$$
$$R_{B_1} = 9.47\ K\Omega$$
$$R_{B_2} = 10.26\ K\Omega$$

There are templates for some symbologies, and the templates will create a neat product created by hand. IBM Selectric typewriters have elements, one of which is for math. These are the "golf balls" of print that got the job done before word processors improved matters. With a Selectric you can create typed calculations. Now, of course, computers will do a sensational job. You will find equation editors on Microsoft Word (as early as Version 5 for the Mac and Version 2 for Windows). You can call up the equation and then import math into your document. Any of these three options is acceptable: longhand, typing, or processing.

A calculation can also be integrated into a line of type in the text. Use this method only if the mathematical logic is simple and the calculation is easily structured on a line. The most trouble-free method of presenting calculations is to display them by placing the calculations in the center of the page.

---

3. Data lines A, B, D have high levels (1) on the middle bit position corresponding to parity bit number 2, or

P2=A ex-or B ex-or D

4. Data lines A, B, C have high levels (1) on the left bit position corresponding to parity bit number 3, or

P3=A ex-or B ex-or C

For the parity generator, the exclusive-or gate was chosen because its outcome on the Karnaugh Map showed high states only in diagonal boxes. This requirement is met specifically by the exclusive-or gate.

You can also place calculations near the left margin. To do so, indent five spaces and begin the calculations:

It should be apparent that Equation 1 is equivalent to saying

$$3478 = 3 \times 10^3 + 4 \times 10^2 + 7 \times 10^1 + 8 \times 10^0.$$

The general form is

$$X = a_a \times 10^a + \ldots + a_2 \times 10^2 + a_1 \times 10^1 + a_0 \times 10^0$$

for all numbers in the decadic numbering system.

Both systems are common. Whichever one you choose, use it consistently for the length of the document.

The use of the calculation display is helpful, but there are other assists that will ease the reader through the work. If there are few calculations, you need not number them. If, however, you need to refer your reader to calculations, number all the calculations—or at least the ones you refer to (actually, entire groups of calculations, theorems, and equations and not every line). Number displayed calculations only. Do not number calculations printed on a typed line. For example,

---

Thus, for a capacitor C used in a sine wave application at frequency f, we have

$$\boxed{\text{Power loss} = 1/2 \; Cv^2 \times DF \times f} \qquad (3)$$

where v = rms values of the sine wave.

The preceding layout now becomes a handy reference to the third calculation. If you place all the calculations near the left margin, you can place the reference numbers on the left also, although this practice is less popular. Of course, you do not want to confuse reference numbers with the calculations, so indent the reference number five spaces from the left margin, and then indent the calculation at least another four or five or more spaces to separate it from the reference. The reference number can also be titled, as *Equation 1*, for example.

Recalling the equations of electric F_e and magnetic F_m fields and the principle of vector field superposition,

1) $\vec{F}_e = q\vec{E}$ (force due to the electric field = F_e)

2) $\vec{F}_m = q(\vec{v} \times \vec{B})$ (force due to the magnetic field = F_m)

3) $\vec{F}_e + m = F_e + F_m = q[\vec{E} + (\vec{v} \times \vec{B} B)]$ (Lorentz Force)

where q = change of the ion

$\vec{v}$ = velocity of the ion,

it is necessary to employ the two vector fields (equations 1 and 2) in such a way that when equation 3 is applied to the ion stream the resulting force on the ions produces the desired effect.

Symmetry is an element of clarity, and a few guidelines are standard practices for math. If your discussion begins with any of the following mathematical terms, print the statement after it in italics.

Definitions

Theorems

Lemmas

Assumptions

Rules

Use these terms as on-line headings prior to the statement. This practice results in an example such as you see for the following theorem from the American Mathematical Society.

THEOREM 6. *There exists one and only one degree function in the extended sense on the class F of maps f: cl(G) → X*, where X is a reflexive Banach space and the maps f are pseudo-monotone, with the degree invariant under affine homotopies and normalized by the duality mapping J.*

Another example follows.

1.3. THEOREM. *Assume $\phi \in$ Aut($\pi_1(Y)$), where Y is a finite graph with one vertex. Assume f, f': X → Y, v ∈ V(X) satisfy G1–G4 in 1.2, with f' * of *$^{-1}$ = ϕ. Then f* (ε_v) ≅ Fix(ϕ) is finitely generated.*

The calculations then follow, always by their own logic, so that, for example, conditions or equivalencies will always follow equations and so on. Here I have borrowed another example from the Society.

Usually, as in the preceding example, the calculations are symmetrically aligned, most commonly by aligning a symbol or symbols on the left of each line of calculations. *The equal sign is particularly effective for giving a reader a sense of symmetry.* I say sense of symmetry because the math is read to the right and not vertically. The neatness has no real value except in the columnar math of the business world. Equivalencies are often part of the display, and they are particularly easy to handle because of the equal sign. Place the word *where* to the left, and neatly present the equivalences below the calculations they explain.

$$
\begin{aligned}
L &= L(V,C) && \tag{5}\\
&= C^a (V - V_o)^B && ;\text{for } V > V_o\\
&= 0 && ;\text{for } V >= V_o
\end{aligned}
$$

where:

V	=	Vulnerability factor
V_o	=	Threshold vulnerability level
C	=	Vehicular capacity ratio
	=	$(D\,N)/(D\,N)_{min}$ where $(D\,N)_{min} = 1.0$
DN	=	Product of total bridge length, D, and number of lanes, N. This number reflects the total number of vehicles that can be accommodated on the bridge. The subscript "min" refers to the smallest product value in the group.
a, B	=	Exponents evaluated below

If the calculations exceed the length of the displayed width, they can be continued on the next line in much the same manner as conventional text lines. The break point should be logical. The equal sign is probably the ideal break. Otherwise, stop at an operational or function sign but *not* one inside parentheses or brackets. The order of preference is at the plus sign first, the minus sign second—or at the multiplication sign if there are no other options. (There are three multiplication notations: the x, the midline bullet or multiplication dot, and the asterisk, which is used in program languages.) For example, here is a calculation that is used for bending stress analysis.

$$\sigma_{zz} = Mx[(I_{yy}y - I_{xy}X)/(I_{xx}I_{yy} - I_{xy}^2)] + M_y[(I_{xx}X - I_{xy}y)/(I_{xx}I_{yy} - I_{xy}^2)]$$

The equation can be divided. There is no practical location near the middle with an an equal sign, and I cannot divide the parenthetical component, so I use the plus sign as the break. The sign moves to the next line with the remainder of the calculation.

$$\sigma_{zz} = M_x[(I_{yy}y - I_{xy}X)/(I_{xx}I_{yy} - I_{xy}^2)]$$
$$+ M_y[(I_{xx}X - I_{xy}y)/(I_{xx}I_{yy} - I_{xy}^2)].$$

There is a major final consideration: continuity. The discussion so far has concerned a neat, orderly, and logical format that you can use consistently. This much will assure the continuity of the *design* work. The *logical* continuity is a little more complicated. Remember, you are expecting a reader to shift back and forth between several symbol systems at once. Your obligation to the reader is to make the movement occur without any jolts. The goal is an orderly and smoothly managed progression.

The general principle to be aware of is *flow*. Make the sentence move smoothly by using English as the cue. Use indicators (for example, *this equation* . . .) and transitions (for example, *then*) and punctuation (for example, the colon) and layout. Although there is no absolute rule for placing punctuation *after* the calculations themselves, you can do so to help the reader see the sentence organization. I'm always asked what to do if a sentence ends on a calculation, since a new sentence might not be clearly indicated to the reader. Grammar rules have always discouraged the use of numbers to *start* a sentence, since there is no capitalized word to tell the reader that it is a new sentence. By that logic, you should not end on a calculation—unless you put a period after it, which you can indeed do. The only argument against such a practice would be that the period or comma might be confused for part of the calculation. You must select which point of confusion—the sentence continuity or the calculation continuity—you want to address. You definitely should not add a period after an Internet address or after a data processing or other display where the period could cause problems.

Calculations are often pieced into sentences, as in the following sample from a college project. Notice how nicely the continuity is sustained between the two systems. Awareness is the key to success in this framework.

From equation 2, it is clear that the force $\vec{F}_B$ on the ions of velocity $\vec{v}$ will have a scalar value of qvB (if the magnetic field is perpendicular to $\vec{v}$) and a direction that will always be perpendicular to $\vec{v}$. Whenever a particle moves under the influence of a force acting perpendicular to the particle velocity vector, the particle executes circular motion. Recalling Newton's Law of motion

5) $\vec{F} = m\vec{a}$

and substituting in the values of acceleration and force in terms of the particle's velocity, which was established by the velocity selection

6) $qvB = \dfrac{mv^2}{r}$

where r is the radius of the circular path, canceling one v, the equation becomes

7) $\dfrac{q}{m} = \dfrac{v}{rB}$

In the preceding sample you will notice that the colon is not used. There may be a comma, a colon, a period, or no punctuation before the display. Do not use the colon simply because there *is* a display. The logic of the sentence determines the punctuation. The colon is used *only* at the end a complete sentence, as in the following sample.

This problem then becomes one of simple algebra:
$$R\bar{z} = \Sigma Wz$$
$$R\bar{z} = (6400\#)(2') + (3600\#)(7')$$
$$\bar{z} = \frac{12800'\# \, 25200'\#}{10000\#}$$
$$\bar{z} = 3.8'$$

The easiest way to handle the display is to write a sentence that explains the display in some way, but end the sentence with a period. Then display the math, as in the following sample.

From Ohm's law we know that voltage equals current times resistance. If we substitute current times resistance for voltage in Equation 1, we get Equation 2.

$A = \dfrac{r_c i_c}{r_c i_c}$ *Equation 2*

If readers have difficulty with the combination of running text and the displays, it is usually because of displays that occur in the middle of a sentence. The longer sample from the college project is well designed and involves more than one display in a sentence. For some writers, this is not easy to accomplish. The simplest way to handle the displays is to use only one display at a time and to place it at the end of a sentence that uses either the period or colon, as in the two shorter samples immediately preceding.

There is an obvious connection between the running text and the symbolic displays, so the logic structures of the two systems must be carefully linked. To get a sense of this linkage, read the following sample and locate the verb in the second sentence.

> Note that the current flow through R_3 is the algebraic sum of I_1 and I_2.
>
> Since these two currents flow through R_3 in different directions,
> $$I_{R3} = I_1 - I_2.$$

Without the equation, the second sentence has no meaning, but, further, without the equation, the sentence is not a complete sentence, because the verb is the equal sign of the equation!

There are a few additional tips. As you have probably noticed, there are occasional inconsistencies between English rules and technical writing practices. For example, numbers under 100 are written out as a general English rule, but *not* in the specific case of technical and engineering work, in which numbers, even the first hundred, are cumbersome in written form. Similarly, some punctuation rules have to be ignored. American English always called for periods and commas to be placed inside quotations. In any document concerning computer program languages, as I noted, authors usually ignore such rules. Computer program language is based on commands that can be misunderstood if the punctuation is regularized. In other words, a period inside a quotation or display could be a gremlin that could sabotage your program.

Equation Editors

Fortunately, there are abundant supports for developing math displays. Equation editors are now popular features of the currently available software applications. Word 7.0 for Windows and Word 97 have equation editors, but they may be a little hard to find. Look under Help and type "equation editor" in the query window. The "graphing calculator" part of the Mac OS does a nice job of formatting many equations. *WordPerfect 5.1* for DOS, *Scientific Workplace*, *Ventura Publisher* (Desk Top Publishing), and Lotus *Word Pro 96* all have an editor. There may also be third-party, add-on editors for *PageMaker*, *Quark Express*, and others. Your workplace or campus computer facilities may have custom installations that vary in what applications they load. TEX and Latex are shareware systems for formatting math and scientific equations, and they are widely used by mathematicians and scientists on all computing platforms. A popular freeware equation editor *Publicon* (from Wolfram, makers of *Mathematica*) can be used on all recent computer operating systems.

Exercises Chapter 8, Section A

The following math needs help. Restructure each sample so that the math is properly displayed.

Sample 1

According to the fundamental law of electricity (2) that was mentioned earlier, an ac voltage ε_2 is therefore induced in the second coil. The voltage ratio of this ideal transformer can be expressed in the equation, $\varepsilon_1/\varepsilon_2 = N_1/N_2$, where ε_1 = voltage induced in the primary, ε_2 = voltage induced in the secondary, N_1 = number of turns on the primary, N_2 = number of turns on the secondary.

Sample 2

But if our black-box machine could raise the weight only 3 ft (s_0) when the effort force is moved 12 ft (s_i), we would have (4) IMA = 12 ft/3 ft = 4. Similarly, the actual mechanical advantage (AMA) is defined as the ratio of the force of the load to the effort force applied: (5) AMA = F_R/F_E. The machine efficiency is defined as follows: (6) Efficiency = actual mechanical advantage divided by ideal mechanical advantage.

Sample 3

The angular amplitude, theta, associated with each OR can be calculated based on the trigonometric relationship $\sin\theta = $ (OP divided by D). Solving for theta $\theta = \sin^{-1}$ (OP divided by D). Theta can be calculated with one command, where each element in the array OR is divided by the scalar, D, and as in (x) is applied as the inverse sine function.

Chemistry, Physics, and other Disciplines

All the techniques we have discussed for developing mathematical sections of a text apply to other scientific disciplines as well. For example, you can see the exact same continuity in the following chemical *equation*. The reaction that follows,

$$Ca_3(PO_4)_2(s) + 3H_2SO_4(aq) \rightarrow 3CaSO_4(s) + 2H_3PO_4(aq)$$

is the fundamental reaction used in the process of making calcium phosphate for fertilizer mix so that the compounds can be absorbed by plants.

Here is a formula for your coffee if you want it low-cal but sweet:

Or do you prefer sugar?

The two preceding symbolic representations of Nutra-Sweet and sucrose, respectively, use similar graphic methods. The reaction could appear in running text, but obviously the structural formulas must appear as displays. As in the case in math, even if the reaction could be part of a typed line, it will be easier for the reader to follow if the reaction is placed in display.

Each scientific discipline develops symbolic systems to meet its own needs. Chemistry uses such symbolic features as subscripts and superscripts—but they may appear *before* the primary symbol.

$$^{14}_{6}C$$

The symbol above is used to identify isotopes: the upper number is the mass number, and the lower number is the atomic number. To those who read chemistry, this particular notation denotes the radionuclide carbon-14 that is measured as a popular determinant of age in dating analysis in anthropology. As noted earlier, chemistry uses symbols to help visualize the text in which the symbols appear. An assortment of reactions and equilibrium reactions are indicated with arrows. The arrow functions in a fashion similar to the equal sign in mathematics. Like word-based languages, the system is read left to right, top down. To the left of the arrow are the reactants, and to the right of the arrow are the products. You can see the reactants, the reaction symbol, and the products in the following molecular equation:

$$K_2CrO_4(aq) + Ba(NO_3)_2(aq) \rightarrow BaCrO_4(s) + 2KNO_3(aq)$$

The electron bonds are graphically represented in structural formulas with single, double, and triple connecting lines:

Benzene

The result is a graphic element that reminds one of ball-and-stick modeling.

The symmetry of chemistry then moves to a level at which human genius exercises its ideas with Tinkertoys. Here is a typical ball-and-stick conceptualization of a protein chain called the α-helix. If you refer to the structural formula for sugar, you can see that ball-and-stick graphics do indeed visualize the world of the biochemist.

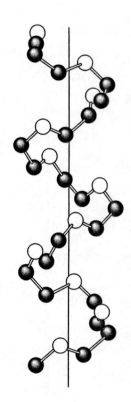

There is an interesting progression from the abbreviations of chemistry notation through the structural formulas to the ball-and-stick modeling; it is a movement to more graphic mediums, although the graphic mediums remain highly symbolic.

The reactions and other equations that are used in chemistry are handled in much the same manner as you observed in math. The easy approach is to explain or describe the display in a sentence that ends with a period.

> This comes in handy if we already know the hydroxide ion concentration of the solution because together the pH and the pOH of a solution will add to 14 (at standard temperature).
>
> $$pOH + pH = 14$$

If you are confident in your skills, you may use the displays in the middle of a sentence, but you must be careful to maintain the logical continuity of the sentence.

We'll start by considering a buffer solution containing a weak acid HA and its conjugate base A:

$$HA_{(aq)} \rightleftharpoons H^+_{(aq)} + A^-_{(aq)}$$

The ionizaton constant K_a of this acid at equilibrium is

$$K_a = \frac{[H^+][A^-]}{HA} \tag{2}$$

Rearranging equation (2),

$$[H^+] = K_a \frac{[HA]}{[A]} \tag{3}$$

and taking the negative logarithm of both sides, we get

$$-\log[H^+] = -\log K^a - \log\frac{[HA]}{[A^-]}$$

or

$$pH = pK^a + \log\frac{[A^-]}{[HA]} \tag{4}$$

Equation (4) is referred to as the Henderson-Hasselbalch equation.

The structural formulas are somewhat of a mixture. They are formulas on the one hand, but they can be quite graphic. As a result they can be handled in two ways. Structural formulas can be placed in simple displays as though they were reactions. The following sample does, in fact, involve a reaction, but notice the captions that identify the compounds.

Acid anhydride is an important class of organic compounds derived from acids via formal inter-molecular dehydration. In most cases, two carboxylic acids of the same kind react together to produce acid anhydride. The product is twice as reactive as the reactants.

O
‖
R–C–OH   +   HO–C–R'   →   R–C–O–C–R'   +   H₂O
Carboxylic acid   Carboxylic acid   Acid anhydride

Structural formulas can also be handled as a graphic, in which case they are presented with a figure number and a caption and perhaps a figure box.

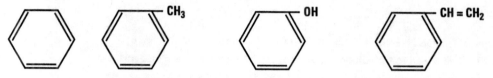

Figure 5: From the left; benzene, toluene, phenol, and styrene

Physics abides by exactly the same display standards that are applied to math and chemistry. I will not repeat a third set of samples, since the procedures are identical, but I will include a typical display in physics. Observe that it is exactly in the style of the procedures you have seen for math and chemistry displays. The following sample concerns the Doppler shift.

At normal incidence if f is the frequency of the incident wave and v is the velocity of the re-flecting boundary toward the source, the change in the frequency due to Doppler shift (Δf) is given by

$$\Delta f = 2\ vf/c \text{ where } v \ll c \ (v \text{ is much smaller than } c)$$

This chapter has taken a practical look at how to handle mathematics and chemistry in technical presentations.* Whether you look at astronomy, physics, civil engineering, mechanical engineering, or any other scientific or engineering discipline, you will always see the need for a written text that embraces both the languages of the sciences as well as the language of the spoken tongue. This chapter concludes with a well-designed student sample on the next page. Notice the way layout and symmetry lend clarity to the project. Also, note that the written part of the text is designed to provide continuity for the symbolic elements of the discussion.

---

* The book I find to be the most convenient source for the basic rules of usage in mathematics and chemistry is The Chicago Manual of Style, 14th ed. (the University of Chicago Press).

Sample 8.A

The following calculation illustrates, quantitatively, the small change in pH that occurs when 0.10 M HCl is added to the solution, assuming a 1.0 L total volume. Since this is a buffer solution, the conjugate base, C^{N-}, will react with the acid to keep the pH constant according to the following reaction:

$$CN^-_{(aq)} + H^+_{(aq)} \rightleftharpoons HCN_{(aq)}$$

Initially we have

$$[CN^-] = 1.24M$$
$$[HCN] = 1.24 \text{ M}$$
$$[H^+] = 0.10 \text{ M}$$

According to the above equation, the molar ratio of H^+ to CN^- and to HCN is 1:1 in both cases, resulting in the loss of 0.10 mole of CN^- and the gain of 0.10 mole of HCN. This can be summarized as follows:

	$CN^-_{(aq)}$ +	$H^+_{(aq)} \rightleftharpoons$	$HCN_{(aq)}$
$[\ ]_0$:	1.24	0.10	1.24
change:	−0.10	−0.10	+0.10
$[\ ]_e$:	1.14 M	0M	1.34 M

Using the Henderson-Hasselbalch equation, we obtain the new pH for this buffered solution:

$$pH = -\log(4.9 \times 10^{-10}) + \log\frac{[1.14]}{[1.34]}$$

$$= 9.24$$

Obviously, the pH does not change significantly (from 9.31 to 9.24) in the buffered solution.

The two preceding examples demonstrate how effectively the buffer does its job.

Exercises Chapter 8, Section B

The following chemistry explanations are confusing. Restructure each sample so that the reactions and equations are properly displayed.

Sample 1

In a single replacement reaction, one element in a compound is replaced by another element. The reaction has the general formula AB + C = AC + B. To show an example of this we could use $Zn + CuSO_4 = Z_nSO_4 + Cu$. In this example, we can see that the zinc and the copper atoms exchange places with each other. Also visible in this reaction is the fact that the zinc atom has lost two electrons. The copper ion gained them.

Sample 2

For the reaction, the rate constants are the same. Dividing Equation 4 by Equation 3 means that the rate constants, k_1 and k_2, can be canceled and gives the ratios $Rate_2/Rate_1 = [SO_2]_2 [H_2]_2/[SO_2]_1 [H_2]_1$ (Eq.5). Because the concentration of H_2 was the same, we can cancel these two terms, and the equation becomes $Rate_2/Rate_1 = [SO_2]_2/[SO_2]_1$ (Eq.6). Substituting the data from Table 1 into Equation 8 gives the following: 9.90×10^{-7} $M \cdot s^1/4.95 \times 10^{-7}$ $M \cdot s^{-1} = (3.00 \times 10{-}3\ M)/(1.50 \times 10{-}3\ M)$ (Eq. 7).

Reorganize the following discussion of electronics so that the discussion of Kirchhoff's law is clearly and properly displayed in the sample.

Sample 3

Kirchhoff's voltage law can be applied to the base-emitter loop of the transistor Q1 to determine the current through the base. The current through the collector and the current through the emitter are approximately the same. When the transistor is turned on, there is a very small voltage drop across the collector-emitter. In this analysis, we will assume that there is no voltage drop: (1) $-R_{in}(I_b) - V_{be} - R_e(2{*}I_e) + V_{ee} = 0$ and (2) $I_b = I_e$ divided by β_{dc}. We also know that (3) $I_c = I_e$. Therefore, the current through the base can be derived from equations (1) and (2): (4)$I_e = V_{ee} - V_{be}/2 {*} R_o + (R_{in}/\beta_{dc})$. $V_{be} = 0.6$ V for a silicon transistor and 0.2 for a germanium transistor.

CHAPTER 8 SYMBOLIC SYSTEMS: THE LANGUAGES OF SCIENCE

Software Assistance

Men and women in engineering technology programs developed many of the samples in this chapter. If some of the presentations look professional, it is not so much because the publisher of this book has tampered with the products but because the software that can support engineering and technical documentation is now widely available. Unfortunately, the products are also little known. Popular supports such as Word or Page-Maker or Windows are familiar to most college students, but these programs may not do the job. Highly dedicated software is, on the other hand, less common. Following is a list of a number of applications-oriented programs that may be of value. Whether you need a drawing routine for chemistry or a schematic design program for electrical engineering, you will find that there is an increasing availability of material dedicated to your specialized interests.

If you inquire about these programs, be sure to ask for student versions. The application will probably meet your needs, and the difference between student and professional versions can be $1000.

General Science
    LabView, National Instruments   (512) 794-0100

General Engineering
    Origin, MicroCal Software   (800) 969-7720

Anatomy and Physiology
    A.D.A.M., A.D.A.M. Software Inc.   (800) 755-2326

Chemistry
    Beaker, International Thompson Publishing
        (for Brooks / Cole Publishing)   (800) 354-9706
    ChemDraw, Cambridge Soft
        (formerly Cambridge Scientific Computing)   (617) 491-2200
    CSC Chemoffice, Cambridge Soft   (617) 491-2200

Computer Service
    VEDIT, Greenview Data   (800) 458-3348

Construction Engineering
    AutoCAD, Autodesk   (800) 964-6432

Electrical Engineering
    PADS, PADS Software   (800) 554-7253

PSPICE
Cadence Design Systems   www.orcad.com
ICAP for Windows   (310) 833-0710

Electronics
Circuitmaker, Protel Technology   (800) 419-4242
Electronic Workbench,
        Interactive Image Technologies   (416) 361-0333

Flow Charts
Flow Charting, Patton and Patton   (800) 525-0082
Flow Charting PPQ, Patton and Patton

Mathematics
Mathematica, Wolfram Research   (217) 398-0700
MATLAB, The MATHWORKS   (508) 653-1415

Mechanical Engineering
DesignWorkshop, Artifice   (800) 203-8324
AutoCAD, Autodesk   (800) 964-6432

Physics
Interactive Physics, Knowledge Revolution   (800) 766-6615

Structural Engineering
Design Workshop, Artifice   (800) 203-8324

Various Drawing Programs
Smart Draw, Smart Draw Software Inc.   (858) 549-0314

   • Flowcharts, technical drawings, diagrams, Web
     graphics, business presentations.

Visio Technical, Visio Corp.   (800) 248-4746, Ext. LO 46

   • For network diagrams, database diagrams, HVAC layouts, piping and instru-
     mentation drawings, manufacturing and assembly drawings, fluid power
     schematics, and others.

The Proper Handling of Your Resources

A considerable amount of attention in technical documentation is devoted to resources. The point of origin for a given piece of information or bit of data may be of slight interest—or of great concern. Suppose you are reading an article on bypass heart surgery and balloon angioplasty in the science section of your favorite news magazine. According to the article, patients who undergo such invasive procedures are 2.5 times as likely to be *dead* within a month as heart patients who are treated with less aggressive techniques. You think to yourself, "What kind of medicine is this?!"

Then you think, "Well, maybe this is wrong or maybe this is some kind of sensationalism." Reading on you find that this information came from the most trusted journal among medical journals: the *New England Journal of Medicine. Because* the journal carries enormous weight among medical professionals, this particular article created considerable controversy! The source gained as much attention as the data. To put it another way, the data gained attention partly *because* of the source. On a more cheerful note, if you are reading about the extraordinary powers of vitamin C and the evidence in support of the fabled curative, it might be of interest if the citation turns out to have been derived from Linus Pauling's text *Vitamin C and the Common Cold* or other books by him. Pauling was a very important scientist. He was a winner of the Nobel Prize and one of the great chemists of the twentieth century. His critics in the American Medical Association disparage the vitamin C issue and do not think very highly of Pauling. The jury is still out on this question, but you can be certain that any supporter or detractor of the vitamin C issue will want to know the sources for the data.

The ability to evaluate resources is an important skill; how much you need to know about the methods for presenting resources is another issue. In the *Wordworks* series, I have assumed that I have two audiences, the engineering technician who is completing a two-year technical program in an engineering-related program, and a potential engineering transfer student who will move from a two-year program at a junior college or community college to a four-year engineering program. In industry, writing tasks that have to be handled by the two-year graduate are likely to be somewhat different from those that will be encountered by the four-year graduate. The extent to which literature is used as a resource can differ dramatically between corporate writing practices and academic writing practices. In addition, the measure to which the resources are identified—meaning that there are citations for articles and books—can vary dramatically as well.

It is safe to say that a great deal of corporate research is proprietary. There is not much profit to be made from sharing. In the computer industry, for example, all the major movers, such as Apple, IBM, and Microsoft, work under wraps. Indeed, the thousands of industries that have built their empires around patent rights also have an interest in copyrights. The inventions of industry are protected—and all the peripheral paperwork is also. Recently, an aerospace engineer was writing a paper for me that was going to include a detail of a CATIA assembly drawing from a proprietary wing design—until his boss said no.

Industrial competition means that confidentiality and entitlement protect the product—and the profit—which suggests that a corporate employee is not too likely to structure many technical documents in which he or she uses outside resources or, when using them, cares to acknowledge them. You will notice that most documents that emerge from a corporation are *primary;* that is, a document is the point of origin for the material contained within it. Any citation of sources indicates that the document is partly dependent on a secondary resource, and this may be undesirable from a corporate perspective.

Colleges, in contrast, emphasize the teaching of research writing. English departments on every campus make a heavy investment of instructor effort on "Research 101." By one name or another, this generic course on references and resources stretches from coast to coast. Why? Because colleges and universities operate on a somewhat different perception of research. Campuses are known for intense and thorough research standards, and these values are nowhere more apparent than in the gathering and studying of data. Colleges and universities value *secondary* sources as some of their major training tools: books, journals, magazines, and the newer electronic resources such as databases and the Internet.

So there are two worlds out there with which you may have to deal. Engineering technicians go directly into industry with their applications-oriented skills. They may, to a greater or lesser extent, be responsible for writing. In fact, the earlier book, *Basic Composition Skills* argues that increased responsibility usually *means* more writing. However, corporations may not utilize footnotes and bibliographies in their documentation.

The issue is a little more complicated for engineers. Their skills are applied, of course, but their work generally involves more analysis, particularly if they are concerned with research or development during their college years. The college and university experience of future engineers certainly calls for skill in research and proper handling of source citations.

The differences between corporate privacy and academic practices, and the differences between applications-oriented skills and analytical skills, should define your perspective for reading this chapter. Any discipline that calls for college-level education assumes that you are to be able to *read* all the parts of documents with competence—and that includes all manner of citations and bibliographies. There you might draw the line. The need to *write* citations and references is a more academic matter, largely restricted to research of one sort or another.

In other words, all engineering tech college students need to be comfortable with citations and references, but far fewer of them need to devote much time to constructing these tools. Read the material that follows with an eye to understanding *your* needs. The discussion explains the popular practices that are used to develop *citations, bibliographies,* and *text notes.* One reading of the material will make the procedures clear to you so that you will understand how to read document resources. If you want to develop your own citations, bibliographies, or text notes, refer to the specific MLA (Modern Language Association) conventions that identify convenient practices for designing these features for college projects. If your instructor prefers that you follow American Psychological Association standards (another popular style standard for academic writing), then follow your reading of Chapter 9 with a close reading of the appendix (see p. 407).

Source Citations for In-Text Documentation

Let's keep this particular discussion as simple as possible. Entire books are devoted to the craft of documentation; there is one source each for biology, chemistry, electronics, geology, mathematics, medicine, physics, and psychology. Each field has its own specialized documentation style. (Notice that only one of the *engineering* fields is on the list.) You would need only one style manual if all these disciplines used the same methods. Because they do not share standards, this discussion is a little complex, but the analysis will keep the procedure practical and brief. You have only a few basic needs, after all.

1) You need to know how to refer to resource material *in the text* of your document.

2) You need to know how to construct a list of resources *at the end* of the text.

3) You need to look at possible formats for explanatory material *at the bottom of a page* or in *endnotes.*

When you examine these practices, pay close attention to order and punctuation. Documentation is a precise business. It always was, but now that research has taken a turn toward computer-network queries and worldwide interlibrary standards, the accuracy of resource documentation can help your travels on the new information highway.

First, you need to understand in-text acknowledgments. Suppose you are reading a work on the applications of mass spectrometer analysis in geology. In the text the author acknowledges the sources of a number of observations. The author can do so in any of the following ways:

1) Geological age analysis methods have become quite accurate (**Aronson 231**).

2) Aronson is optimistic about new trends in geochronology (**101–118**).

3) Spectrometer analysis according to Aronson (**1989**) is an effective tool in geochronology.

4) The argon analysis indicated that the basin was 2 billion years old (**Aronson**).

5) The argon isotopes are quite accurate for geological dating (**Aronson, 1989**).

6) Heating factors create variances in the isotope density (**Aronson,*Geological Isotope Structures***).

7) Argon accumulates in certain mineral deposits (**2:231**).

8) The argon method is effective (**2**).

9) Under analysis, the geologic age is easily established.[2]

As you can see, at least nine options could show up in documents depending on the preferences of various disciplines. This list is not structured by popularity but by method. Each approach is easily explained.

(1) The first citation is widely used in college and university settings. It is the MLA citation method taught in most English courses. The sentence ends with a name and number in parentheses indicating the author (and page number) from whom the writer has borrowed an observation or information. This is a cue to a reader that Aronson will be included in a list of references at the end of the article. (All the samples are boldfaced to highlight the citations. They are not boldfaced in normal usage.)

(2) The second example, also an MLA convention, is the other popular method of citing a source of information in college papers. In this instance, the author's name is part of the sentence and only the pages are listed in parentheses as a source.

(3) In the third example the year appears in parentheses to identify which one of this author's works is referenced in this sentence.

(4, 5, 6) These examples show other variant forms that identify a source by the author, the author and the date of the document, or the author and the title of the source in parentheses.

(7, 8, 9) If there is only a number in parentheses, you are, of course, being forwarded to a list of sources for the borrowed information or for more information. This list of information will be in a footnote or endnote. The example (2:321) also provides a page number. Samples 8 and 9 are references to the location of the document in a list or in a footnote. The American Association for the Advancement of Science uses the (2) shown in example 8 in their magazine *Science*. The superscript 2 shown in example 9 is a style used in some medical publications.

Which is correct? All of them, so you must determine which one is used in your field. Which ones are the most popular in general usage? Two of the more popular devices are the author-year reference (APA) and the superscript reference. Both samples refer the reader to a list of resources.

> The microwave frequency of the atomic clock is 9, 192, 631, 770 hertz (**Eisen, 1991**).
>
> He questioned whether there was a metric invariant for automorphisms.[3]

Both indicate that data or information has been borrowed or mentioned. The author-year format has the added benefit of indicating points of information for readers when they encounter borrowed material. They see the author and the date of the work. Both facts can be subtle measures of the accuracy of the data.

The other two citation methods that you are most likely to see are the first two on the list of nine: the MLA citations. Pay close attention to these because they may be the most convenient practices for you to adopt in college work.

MLA Citations

Author and page

Geochronology is sufficiently site-specific for spectrometer analysis (**Aronson 231**).

Page reference to an in-text author

Smith explained that diamonds are the best conductors of heat and that they are far superior to metals in conductivity (**411**).

Multiple authors (up to three)

The reported percentage of error in any dating is considered insignificant (**Jennings, Paulson, and McBrickle 301**).

Since the MLA methods are taught in most academic settings, use these citation methods if you have reason to develop source citations in your work.

Consistently use only *one* of the systems in any one document to refer to resources. If you *also* have explanatory notes, use footnote numbers for those notes, and do not use footnotes to give credit for borrowed materials.

Short-term memory (STM)[4] is frequently altered by high doses (**Farland 227**).

An explanation of short-term memory would be at the bottom of the page in a *footnote* or at the end of the document in a list of *endnotes* (again identified by number).

Numerical superscripts (the raised numbers) are often used for both resources *and* notes.

A one-micron crystal[5] has unique oscillation characteristics according to researchers.[6]

Superscripts can cover all the uses in one simple notation method. Nonetheless, the author-page approach to citation is convenient because it gives more information than a reference number does, and the device is so simple to use—even if you might have to add footnotes for explanations.

Notice that there are no quotations in any of the examples. Quotations are not likely to be of concern to you. In engineering fields, authors are more inclined to borrow ideas, data, tables, and other material. These sources are *also* acknowledged. For example, return to the list of the nine citations and look at the first two again. The examples are not quotations. In the first citation, the comment is derived from a single page, suggesting that the sentence is a *paraphrase,* meaning that the author has not quoted Aronson. Instead, he has loosely restated the borrowed concept or ideas. The second sample is obviously not a quote or paraphrase, since it is derived from many pages in Aronson's text. It is an *overview* or general comment about a large segment of material. These two techniques of borrowing material are likely to be of more use than direct quotations.

To acknowledge the source of graphics, the popular practice among publishers is to place a *credit line* below the illustration. It is generally understood that college authors have not sought legal permission for their borrowed resources, and the popular convention of identifying an author and a book title should be adequate.

(Courtesy of Harold Brendon, *Kinetic-Molecular Theory*)

An alternative is the expression "Reprinted from."

(Reprinted from Peterson Optics, Inc., Brochure 309-D.)

Because this popular convention does not identify a page number, be sure the source is listed in the bibliography if there is one. If there is no bibliography, you must still credit sources for the graphics because you must respect *fair use of intellectual property.*

Three of the many citation methods are presented next. Note that the first one is the one you would probably use in your college work.

Samples 9.A

1 To increase the capacitance for a given size, d could be reduced and, by using more layers, A increased. Both reducing d and increasing A will increase the number of flux lines through the dielectric. This allows the field to store more charge in the dielectric. This process, however, is limited by the maximum field strength rating *(breakdown voltage)* of the dielectric, since, for a given voltage, the electrical field increases as we decrease the thickness; S is constant and can't be controlled, but K, the dielectric constant, can range from one to several thousand for various insulating materials and is, therefore, an important parameter. Almost all capacitance variations in *temperature, frequency,* and applied *voltage* are due to variations in K **(Meiksin 84)**.

2 An overview of the modem and its basic function is first necessary. The term "modem is an acronym for the words MODULATOR-DEMODULATOR and is a device that permits the transmission of digital data over communication links"**(1)**. In other words, if two DTEs, a CRT, and a computer (digital devices), located in separate physical locations, are to communicate via telephone lines (analog), then the modem is the necessary interface device.

3 The lungs in humans are elastic bags or sacs located in the thoracic cavity. The right lung has three lobes or sections, and the left lung has only two, which allows room for the heart **(1:314)**.

Exercises Chapter 9, Section B

Develop a page of text in conventional double-space format. Use three or four resources that have to be identified because you borrowed data, tables, graphics, or some other material. In each case, after presenting the source material, enter a parenthetical source citation.

If your instructor requires MLA style citations, cite the author and the page number in parentheses. If your instructor requests the APA style (see the appendix), cite the author and the year of publication.

Your page will look similar to the following sample:

Frequency Response

Passive RC filter circuits use discrete components—resistors and capacitors—combined to make a frequency-select circuit. To determine the frequency you want to pass to the output of a circuit, you must determine the values of R and C (**Dunn 316**).

Capacitive Reactance

Capacitive reactance is the output result when an AC (alternating current) signal is applied across a capacitor, and the capacitor charges or discharges in relation to the output load resistor. The formula for capacitive reactance is

$$Xc = \frac{1}{2pfc}$$

where:
$$Xc = \text{capacitive reactance}$$
$$p = 3.1415$$
$$f = \text{frequency}$$
$$c = \text{capacitance}$$

From this formula (**Kraska 74**) you can see that as the frequency increases, the capacitive reactance decreases.

The Reference List for After-Text Documentation

Once you have adopted a method of identifying sources *in* the text, you have to develop a list of those materials at the *end* of the text. Bibliographies can reach massive proportions, particularly if they are designed to acquaint a reader with the *sum* of available material. The only bibliography of real concern to you, though, will usually be the job-specific sort that identifies only the immediate resources that you used for a project. The focus of any particular piece of research is usually quite narrow. List *only* the material you use. Compounding the resources is a college version of *data dumping* or adding unnecessary information.

The bibliography is simply the directory of your resources. The list identifies the authors you consulted for your data. The bibliography is frequently called *References* or *Works Cited* or *List of Works Cited,* and so on. Unlike the many variations of citations and references *to* the bibliography, the bibliography itself has few variables. For all practical purposes, there is only one way to design a bibliography.

Layout

The bibliography is the last document in your text. It is placed after the text and also after the glossary, if you have one. On a separate but numbered page, the bibliography is identified at the top center, and the resources are listed, in alphabetical order, by the authors' last names. The secrets to a well-designed list of sources are few.

1. Double-space each entry, and double-space between them if you are using MLA style. You will often see bibliographies in single-space format. Remember, a college project is a manuscript and it is handled differently than a published paper. Once the product is published, the double-space format is changed to single spacing.

2. Use the first line as the left margin, and indent *subsequent* lines five spaces. This is exactly the opposite of the method of indenting paragraphs, but the first line begins on the left margin so that all the authors in the alphabetical list are easy to identify. Notice the two bibliographies in Samples 9.B and 9.C.

Detailing the Entry

There are dozens of different resources that could be documented, but there are perhaps a half dozen that constitute, by far, the greatest volume of bibliographic space in technical documentation.

- a book
- an article in a magazine
- an article in a journal
- an article in an anthology
- an electronic source
- a corporate document

In its most basic form, the reference to a resource is divided into five parts:

1) **Author(s)**

2) **Title**

3) **Location of publisher**

4) **Publisher**

5) **Date (of your copy)**

Because the structure is a typed line, the five parts have to be punctuated. Here is the generic form for a book entry:

> **Doe, John. <u>Title.</u> City: Publisher, Date.**

1) **Author.** Last name first—comma—first name—period.

2) <u>**Title.**</u> Full title of book—underlined or in italics—period.

3) **City.** City—colon. Add state or country if not well known.

   With the state included, the order is city—comma— state abbreviation or country—colon.

4) **Publisher.** The publisher—comma. Such words as *company* or *corp* are often left out.

5) **Date.** Add the most recent date you see in the copy you are using— and a period.

Every entry is a variation of this basic logic. For technical writing, there is a specific reason for always referring to this basic concept because you will often have to use the prototype to construct bibliographic entries. Corporate documents *often* lack one or more of the standard details just identified, and you will have to contrive a substitute by following the *basic* logic of the entry pattern. For example, there is seldom an author in corporate documentation; there may not be a city; there *is* no publisher except the company itself; dates are often omitted (intentionally), and the document may have no clear title. As a result of these problems, the corporate document calls for your attention, since the material, regardless of its source, must fit into the bibliographic logic structure. Online databases and other materials accessed from computer networks must similarly be adapted to the basic format.

Because entire volumes are devoted to documentation practices in several scientific disciplines (see p. 314–315), you can consult one or more of them for their recommendations when you want to indicate unusual resources: from Edison rolls to CDs, from telephone calls to encyclopedias, from rock records to sheet music, from Mayan codices to CompuServe services. These style manuals provide recommended samples for almost everything from newspapers to lecture notes to electronic journals.

Let's look at samples of the six types of documentation that are most common in technical work. The first sample is a typical book entry format.

Book

Name. <u>Title.</u> City:Publisher, Date.

Sample Application:

    Mcneill, Daniel, and Paul Freiberger. *Fuzzy Logic: The Discovery of a Revolutionary*

        Computer Technology and How It Is Changing Our World. New York: Simon

        & Schuster, 1993.

The turnaround time for scientific and engineering papers is a reflection of the pace of scientific and engineering production. Book publishing is much too slow to keep up with the state of any rapidly advancing field such as digital electronics. Book publishing takes more than days and more than months. Major publishing activities involve years. The information highway may speed up the dynamics of access a little, but fast access has always been around—in magazines and journals. Because they are periodic and because less production time is needed, magazines and journals are your best current source. Many of the available online databases are, in fact, journals and magazines.

Magazine Article

Name. "Article." <u>Magazine</u> date: Pages.

Sample Application:

    Reed, Mark. "Quantum Dots." *Scientific American* Jan. 1993:118–23.

Professional trade journals and the journals of professional societies are usually the most specialized sources available (see pp. 309–312). By all means, consider a membership in the IEEE (for electrical engineers) or similar groups. Professional societies often offer student rates for membership (which includes the journal of the society). Notice in the following example that journal articles include the volume number in case the entire year consists of one set of page numbers. Libraries usually bind together a year's worth of this type of journal.

Journal Article

Name. "Article." <u>Journal</u> Volume (Date): Pages.

Sample Application:

Veech, William. "Dynamics over Teichmüler Space." *Bulletin of the American Mathematical Society* 14 (1986):103–106.

Professional trade journals are less scholarly than society journals and have a strong industry orientation. Trade journals are also very important sources of timely technical information. Although they are referred to as trade *journals,* they are usually handled like a magazine in a bibliography.

Another documentation entry is one that identifies an essay in an anthology (a collection of essays in one book). This entry may be of particular use in science and engineering areas. I noted that publishing can be a slow process that cannot keep up with advancements in scientific fields. There is a compromise. Editors can gather up articles and reprint them quickly in anthologies. The entry illustrated next is a reference to an essay in an anthology. Many magazines, such as *Scientific American,* publish their own anthologies, and they can do so rapidly and efficiently.

Article in an Anthology

Name. "Article." <u>Book.</u> Ed. Name. City: Publisher, Date. Pages.

Sample Application:

Tuan, Yi-Fu. "Topophilia and Environment." *The World of Science.* Ed. Gladys Leithauser and Marilynn Bell. New York: Holt, Rinehart & Winston, 1987. 423–428.

If there is a need to identify electronic sources, the format of the bibliographic entry is an adaptation of the hard-copy samples already examined. The original points of information for hard-copy documentation come first, followed by the relevant supplemental information for electronic retrieval. Although a number of electronic sources can be identified, your most likely references will involve CD-ROMs or whatever material you obtain from the Internet. Both sources begin with the basics and add either technical identification or retrieval information.

Hard-copy text		CD-ROM source
1. Author(s)	↔	1. Author(s) if known
2. Title (underline)	↔	2. Title (and version)
		3. Medium (CD-ROM) and version
3. Location of publisher	↔	4. Location of manufacturer
4. Publisher	↔	5. Manufacturer
5. Date (of your copy)	↔	6. Date of release

Hard-copy journal		Online periodical
1. Author(s)	↔	1. Author(s) if known
2. Article	↔	2. Title (in quotes)
3. Magazine	↔	3. Database/Periodical
4a. Volume	↔	4a. Volume or number
4b. Date	↔	4b. Date
5. Inclusive pages	↔	5. Inclusive pages
		6. Access date
		7. <URL>

The pattern of information is consistent with the hard-copy formats. A number of other formats for electronic sources are described in the documentation guides identified at the end of this chapter. For online material use any one of the standard reference samples as your guideline; then include the access date and the location number and place the URL in angle brackets. Include the access mode identifier *(http, ftp, gopher, telnet, news)*.

Electronic Source/ CD-ROM

Author. <u>Title.</u> CD-ROM. Version. Location: Publisher. Date.

Sample Application:

Time Table of History: Science and Innovation. CD-ROM. Macintosh

version. Novato, CA: The Software Toolworks. 1991.

Electronic Source/ Online

**Author. "Title." <u>Database/Periodical</u> Volume
(Publication Date) Access date<URL>.**

Sample Application:

> Ostfeld, Richard. "The Ecology of Lyme Disease Risk." *American Scientist*
>
> 36 (July–Aug. 1997) 14 Aug. 1997 <http://www.amscl.org/amsci/>.

When using an Internet address from a written document, do not be surprised if your computer cannot find the site. Specific Web pages (usually found at the end of the URLs) change often. On the other hand, if you are only given a domain (the home page site near the beginning of the URL), the author is not taking you directly to the material in question. This is a catch-22 of sorts. By analogy, a domain is rather like a reference to an entire book. The domain site is more reliable than the Web page reference, but it should probably begin with the words "Search under" to be perfectly clear. You will see this convention but it is not MLA or APA in style.

Do be aware of the limitations of the Internet when you are developing a project. It can be used to access major university library systems, but the millions of hard-copy documents are not available to you unless you are an attending student at the school that has the resources you need. In the corporate sector, most of the significant information of interest to an engineer is likely to be protected by various electronic barriers such as firewalls, and access is limited to those who have confidential access codes. In particular, if you are working on college papers, do not hold up a project with the expectation that information from the Net is forthcoming.

A Corporate Pamphlet

**Author. <u>Title.</u> City: Corporation, Date.
(Some of this information is commonly unavailable.)**

Sample Application:

> *Service Bulletin 430B* (Cleaning Procedure). Supplement to the *710x/720x Series*
>
> *Technical Service Manual.* Alaris Medical Systems, Aug. 1999, 6 pages.

As I mentioned earlier, corporate documents are often difficult to deal with because they are unconventional. The actual document used for the preceding entry is illustrated at the end of this discussion. You can see that there is very little information that conforms to the conventional bibliography style. At least the Alaris sample contains a date, which is often omitted or coded on corporate documents. Many corporations code the date in fine print or exclude it as a marketing procedure so that no one will think the shelf life has expired on their goods. Look for the date at the bottom of the last page, where it is often concealed in a longer document number. In the preceding sample note that the title of the document is usually used to alphabetize the entry in the bibliography because there is seldom an acknowledged author. References to larger corporate documents are often organized so that the corporation is in the author position.

A Corporate Author

Corporation. <u>Title.</u> City: Publisher, Date.
(Some of this information is commonly unavailable.)

Sample Application:
Quatech Inc. *1999 Handbook*. Akron, OH: Quatech. 1999.

Corporate Web sites are quickly becoming popular sources of useful information. The corporate Web page can be just as confusing as the typical corporate document. Again, there is usually no author or original date of posting. Perhaps the city is not identified, and there may not even be a document title!

> *Small Business Portable: Solo 5100XL.* Gateway 2000 Inc. (1998). 6 Mar. 1998 <http://www.gateway.com/smbus/portables/ s5100xl.html>.

> Siemans Hearing Instruments. 31 May 1998 <http://www.siemens-hearing.com>.

There are two examples of bibliographies on pp. 294 and 295. They conform to the standards for the entries that you see in the preceding samples. If you need to develop a bibliography, use the samples as your guide, and use the bibliographies (Samples 9.B and 9C) for the suggested layout of the finished product. *Manuscript* bibliographies are double-spaced. *Published* bibliographies are single-spaced. Therefore, lists of references are double-spaced in college work, but if you are self-publishing, you can single-space each entry. Published bibliographies are easy to read if there is a double space between each single-spaced entry. The samples reflect both conventions.

The pamphlet in figure 9-1 was used to construct the corporate pamphlet reference on p. 291.

ALARIS™
Medical Systems

10221 Wateridge Circle
San Diego, CA 92121, U.S.A.
P.O. Box 85335, San Diego, CA 92186-5335

Service Bulletin 430B

P/N 145093A

Service Bulletins are supplements to ALARIS Medical Technical Service/Maintenance Manuals.

Models Affected: Signature Edition® Volumetric Pump; 7000, 7100/7200, 7101/7201
Date: August 1999
Subject: Cleaning Procedure

☞ *This replaces Service Bulletin 430A, to include a step to verify the proper orientation of the flow control actuator pin.*

Purpose

The purpose of this bulletin is to provide Biomedical Technicians:

1. an updated cleaning procedure, and

2. a step to verify the proper orientation of the flow control actuator pin during the *Preventive Maintenance Inspections* and *Test and Calibration*.

Explanation

Alcohol, ammonia, acetone, benzene, phosphoric acid, xylene, and similar solvents can erode (wear away, pit) or otherwise damage the cam followers and other surfaces of the instrument. The cam followers must be kept clean using a solution of warm water and a mild non-abrasive detergent, and inspected for possible erosion.

The RS232 communication port cover must be closed and securely in place to reduce the potential for fluid to enter the instrument during the cleaning process.

Correct operation of the mechanism latch needs to be verified. This can be accomplished by verifying the proper orientation of the flow control actuator pin.

References

710X/720X Series Technical Service Manual (*identified as P/N 142466; ordered as P/N 141776*)
Service Bulletin 428A (*or more current*), Level of Testing Guidelines and Mechanism Springs

Parts and Tools Required

Not applicable.

Parts Ordering: Refer to Illustrated Parts Breakdown chapter of the Technical Service/Maintenance Manual
Technical Inquiries: 1-800-854-7128, 1-858-458-8003, FAX 1-858-458-7507

Page 1 of 6

Figure 9-1 Corporate Pamphlet (courtesy of Alaris Medical Systems).

Samples 9.B

Works Cited

Ainsworth, Susan. "Soaps and Detergents." <u>C&EN: Chemical and Engineering News</u> 5 June, 1995: 32–54.

Brady, James, and John Holum. <u>Chemistry: The Study of Matter and Its Changes.</u> New York: John Wiley and Sons, 1993.

Cheng, E., and J. Chow. <u>Chemistry: A Modern View.</u> 2nd ed. Hong Kong: Wilson Publications, 1989.

Glasstone, Samuel. <u>Inner Space: The Structure of the Atom.</u> Washington: U.S. Energy Research and Development Administration, 1972.

Hairston, Deborah W. "Solvent Alternatives." <u>Chemical Engineering</u> Feb. 1997: 55–58.

Kneen, W. R., M. W. Rogers, and P. Simpson. <u>Chemistry: Facts, Patterns, and Principles.</u> London: Addison-Wesley, 1972.

Lister, Ted, and Janet Renshaw. <u>Understanding Chemistry: Advanced Level.</u> London: Stanley Rhornes, 1991.

Newell, Sydney. <u>Chemistry: an Introduction.</u> Boston: Little Brown, 1977.

Schell, Dave. "Developing a Pollution Prevention Plan." <u>Environmental Protection</u> Apr. 1997: 21–23.

Samples 9.C

Works Cited

Allard, Louis, Guy Cloutier, and Louis-Gilles Durand. "Characterization of Blood Flow Turbulence with Pulse." *Journal of Biomedical Engineering 118* (1996): 318.

Dvonch, Louis A. *Prevention of Stroke, Heart Attack, Alzheimer's-Doppler.* 21 Jan. 1999 <http://www.webcom.com/ldvonch/doppler.html>.

McDicken, W. N. *Diagnostic Ultrasonics.* New York: Wiley, 1976.

McDonagh, D. Brian. "Ultrasound: Unsung Medical Hero." *USA Today* Sept. 1996: 66.

Stark, C, M. Orleans, P. Haverekamp, and J. Murphy. "Short- and Long-term Risks after Exposure to Diagnostic Ultrasound in Utero." *Obstetrics and Gynecology* 63 (1984): 194–196.

Woo, Joseph S. K. "Obstetric Ultrasonography." *BioSites* 11 Feb. 1999 <http://home.khstar.com/-joewoo/joewoo2.html>.

Exercises Chapter 9, Section C

Two bibliographies follow. The first conforms to the MLA standards presented in this chapter. The second conforms to the APA reference list standards in the appendix. There are a number of subtle distinctions between the two systems. You should be familiar with bibliographic or reference styles, but you would use only one of these methods in a document. Ask your instructor which method you should use, then turn to the sample that conforms to the other style. Reconstruct the list of references to conform to the method you would use.

Works Cited: Published MLA Style

Bao, Arthur, John Chan, and Garth German. "Sigma Designs the Leading DVD Playback & MPEG-2 Decoder Company." *Frequently Asked Questions Regarding DVD.* 16 May 1999 <http://www.realmagic.com/dvdfaq.html>.

Baltazar, Henry. "Vendors Divided on DVD Specs." *P.C. Week* 28 June 1999: 83.

"DVD Overview." *DVD Frequently Asked Questions.* 16 May 1999 <http://www.multisource.com.au/realmagc/dvdfaq.html>.

Joly, Christian. "Changing DVD Trends." *Electronic News* 7 June 1999: 32.

Karbo, Michael B. "An Illustrated Guide to Optical Drives (CD-ROMs, DVDs)." *A Complete Illustrated Guide to PC Hardware.* 16 May 1999 <http://www.mkdata.dk/click/module4c2.htm#DVD>.

Stanek, William R. "Taking Style Sheets to the Next Level." *PC Magazine* 1 Dec. 1998: 334.

Taylor, Jim. *DVD Demystifed: The Guidebook for DVD-Video and DVD-ROM.* New York: McGraw Hill. 1998.

White, Ron. "Divx vs. DVD." *PC Computing* July 1990: 200.

References: APA Style in manuscript

Brown, R. (1998). <u>Russ goes surfing with gif.</u> [On-line]. Available: ttp://www.adobe.com/studio/tipstechniques/phssurfgif/main.html

Dinucci, D., Guidoce, M., & Styles, L. (1997). <u>Elements of web design.</u>Berkeley, CA: Peachpit Press.

Lapuck, L. (1996). <u>Designing multimedia: visual guide to multimedia and online graphic design.</u> Berkeley, CA: Peachpit Press.

Lemay, L. (1998). <u>Teach yourself web publishing with html 3.2 in a week.</u> Indianapolis, IN: Sams.Net Publishing.

Adobe Systems Inc. (1998). <u>Preparing graphics for the world wide web.</u> [On-line]. Available: http://www.adobe.com/studio/tipsstechniques/ GIFJPGchart/main.htm/html

Siegel, D. (1996). <u>Creating killer web sites.</u> Indianapolis, IN: Hayden Books.

Simone, L. (1999, August 1) Easy web authoring. <u>PC Magazine 18,</u> 48.

Sullivan, E. (1997, November 14). The right tool for the hypertext job. <u>PC Week 14,</u> 21.

Magazine Citations and Source Lists

I have identified certain practical conventions for in-text and after-text documentation. In-text references vary a good deal, but the after-text source list is fairly cut and dried. There is one additional issue to discuss. It is quite popular, particularly in the case of technical articles of the sort you see in magazines, to omit the alphabetical listing of sources in favor of a *list of resources* that reflects occurrence of *use* in a given article. In other words, many authors omit a bibliography. Instead, they make a numerical list that indicates their sources by order of appearance in their articles. If all the numbers you see in the running text are in order—1, 2, 3, 4, 5—you can be *sure* the author is referring you to a special list of resources that are numbered as they occur.

> Several systemic fertilizers[7] will assist in the control of mites, aphids, or other insects[8] that might damage hybrid tea roses. These compounds can enter the human body upon contact and should be handled with care.[9]

Of course, this system is just as obvious in the references list at the end of such a document because the entries are not alphabetical, and they often cite page numbers. Notice the following sample adapted from an engineering magazine.

[1] Hines, Frederick. *Designing an Energy Efficient Structure,* Washington: General Services Administration. GSA DC 76-3342 (July 1991): 131.

[2] Barnes, Doris. "Post-Design Analysis of Energy Conservation Options for a Multi-Story Building." *National Bureau of Standards Series* 206 (Aug. 1987): 24.

[3] Taliesen was designed to hug the landscape, but energy efficiency was a minor consideration.

[4] Anderson, Carl. "Thermal Performance in Energy Conservation." *ASHRAE Journal* 63 (Sept. 1990): 409.

This system is something of a cross between footnote references and the traditional bibliography. The notes are simply moved to the end and usually mention all the sources. Do not be confused by this practice. *You have no reason to use this system unless you are writing for magazines.* However, you will frequently see the magazine method.

ibering each source in a bibliography is another common practice. If each source has mber, the in-text references can be very simple. Numbers that follow no apparent r easily identify this practice. For example,

> Gene pool preservation is off to a very late start in the preservation of flora **(6)**, and there is an ever-increasing decline of gene resources as a result of hybridization **(2)**.

e is no numerical logic to the numbers above, so you can guess that they must be ref-ces to an alpha-numerical list of sources that will be found in the bibliography.

ably the best method is to practice the rules of your particular engineering discipline; rwise, use whatever systems are *clear* and *simple*. The alpha-numerical system of the pool sample above is quite popular; on the other hand, why number the alphabet? uld have been simple enough to use the names of the authors.

Explanatory Notes

There are two ways to develop an explanation that you want to place outside the body of your text. If you use the author-page method to identify a source in the running text, and an alphabetic list of those sources (a bibliography), you will not need footnotes for sources. Footnotes are numbered or coded references, usually identified by *superscripts* or *asterisks,* that direct a reader to the *foot* of the page. You can use them for text notes. They are references that a reader easily understands. Readers are accustomed to going to the bottom of the page for whatever explanations they need. The text notes usually appear in fine print precisely because they are optional. This is a handy tool for all types of explanatory material.

There are two slightly different ways to indicate footnotes. One is to place an asterisk *after* the word or commentary in question. At the bottom of the page, you provide an explanation, as needed, in single-spaced text.* Double-space twice below the text. See the following brief sample, and then notice the typical text makeup in Sample 9.D. The use of the asterisk is not MLA practice, but it is a traditional tool and it can be useful for self-published materials or in other nonacademic publishing situations.

The accuracy of the comparator depends on its voltage gain, common-mode rejection ratio (CMRR), input offsets, and thermal drifts. High voltage gain requires a smaller difference voltage (hysteresis voltage)* to cause the output to switch between saturation levels. On the other hand, a high CMRR helps reject the common-mode input voltages, such as noise, at the input terminals. To minimize the offset problems, the input offset current and input offset voltage must be negligible; also, the changes in these offsets due to temperature changes should be very slight. Use the specification sheet to pick the correct op-amp for your purpose.

* See glossary for definition.

---

* You can also use somewhat smaller type or italics for footnotes if you are self-publishing, but this is not MLA style.

Sample 9.D

II. MODELING A TYPICAL BUILDING

A specific application could be the typical contemporary low-rise light commercial building that incorporates cost-effective energy efficiency. Buildings like this are commonly retail space, offices, or light manufacturing companies. With tight construction and no open doors, make-up* air volume can be set just slightly greater than the exhaust requirement. The effect of infiltration can be generally disregarded in a tight building. Make-up air volume will be greater than the fresh air ventilation requirements for the occupants, even without a high exhaust requirement, because high-volume fresh air is now required only in contaminated environments. All these factors have significant cumulative effects. Make-up air is often the single greatest load component in a "loose" building.

The peak load that is selected for mechanical equipment to sustain is generally based on a composite peak time that is based on a twenty-year average of weather conditions. However, in any particular building the component load peaks seldom correspond at any particular time. Even so, low-rise buildings of this scale are generally more sensitive to cooling for external loads than to internal heat gains. In a simple calculation, the solar peak is arbitrarily selected, all other loads are assumed to peak, and the hvac system will often be greatly oversized. Consequently, the system will be power hungry, something building users want very much to avoid.

*Fresh, outdoor air is required to "make up" for exhaust requirements and exfiltration through doors, windows, and leaks as well. This air goes to "make up" the total heating and cooling air volume.

If you need several symbols for several footnotes, you typically use the following ones in this order:

*

†

**

††

These are the asterisk, dagger, double asterisk, and double dagger.

If you use symbols, you begin again with a new asterisk on each page. If you use the other method—numbers—you number each note *all the way through* the text from 1 to, perhaps, 17. You can see the benefits of each of these two methods. The symbols are simple and are less likely to become confused with a text full of calculations, which is why they are commonly used in tables. The numbers allow you to list all the notes together if you chose to do so for some reason. A single use of this sort of notation is illustrated next, and a typical full-page illustration of the technique appears in Sample 9.E. The use of numbers for text notes conforms to MLA style and is useful for academic projects. The notes can also be grouped as endnotes after the text if desired.

From the specification sheet[1] for the op-amp in use, find the maximum input offset voltage. Use this value to calculate the value of the resistors you will need. First, choose a value for one resistor. Call this resistor R1. Use the following formula to find the value of the other resistor (R2).

$$V_{ut} = \frac{R1}{R1+R2}(+V_{sat})$$

1. $+V_{sat} = 14V$ from specification sheet
2. $V_{ut} = $ input offset voltage

[1]*See appendix for specification.*

Sample 9.E

The last great advancement in transistor technology is the field-effect transistor. This transistor does not eliminate its predecessor, the BJT, otherwise known as the NPN transistor, but it is much smaller and more efficient. Although this transistor seems ideal, it has two-major drawbacks. These drawbacks are its low gain for amplification[1] and low tolerance to high voltage.[2]

We will first look at the transistor's schematic symbol[3] for the field-effect transistor that is shown in Figure 1. We see by this schematic symbol that this transistor has four terminals: "G," "S," "B," and "D." Each letter denotes a name, which designates each terminal on the schematic. The "S" terminal is the source terminal, the "G" terminal is the gate terminal, the "D" terminal is the drain terminal, and the "B" terminal is the body terminal.

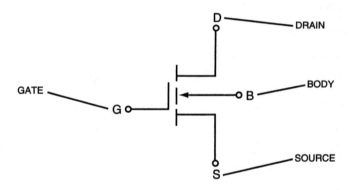

Figure 1 A circuit schematic symbol for the N-channel enhancement-type MOSFET.

---

[1] The process of making a signal larger.
[2] A potential number of electrons that are available across two terminals.
[3] A symbol that represents an electronic part or connection.

Exercises Chapter 9, Section E

Develop a page of text in conventional double-spaced format. Use two to four explanatory notes as footnotes to the page. The explanatory notes used by college students are usually used to explain terminology or, perhaps, concepts, for a specific group of readers.

Use either superscript numbers or symbols (the asterisk and the dagger). If the project is academic in application, use numbers. If, however, the application is work related, the choice of note identifiers is up to you or your employer.

Your paper will look similar to the following sample.

A protocol is a code that specifies what kind of information exchange will occur, that is, how the client (or requesting) computer will access the information from the server. The most common type of protocol used on the World Wide Web is the hypertext transfer protocol (written as http://), which allows the exchange of multimedia information by requesting that only a hypertext document and any elements* embedded in it be sent back to the browser for viewing.

* The term element *refers to a string of text, an image, or any other part of a Web page that is considered as a single entity, but it is not a part of a document's overall architecture.*

Finding Engineering Resources

Libraries

The traditional methods of research involved legwork. A long walk through the endless stacks of a library points to the fact that libraries are one of the best sources of information. For engineering and engineering technology research, you have a number of options that you can pursue to locate material for projects. The *campus library* is the most convenient, not only because you have easy access to the facility but also because the collection of literature will include all the specialty areas that are taught on campus, whether it is electronics, CAD drafting, computer networks, or dozens of others.

The other campus libraries in your area are also prospects to investigate. For example, there may be other community or junior colleges that offer programs similar to the one you are taking. If they are within convenient driving distance, investigate their library collections. They are likely to have somewhat different resources available for your use. Similarly, the four-year colleges and universities in your area are an enormous resource. Major *university libraries* hold millions of books and documents. Call to see whether the nearby universities are open to the public. Many people do not realize that they have free access to state university libraries because they are public institutions (although you usually cannot check material out of the library unless you are an attending university student).

Municipal or county *public libraries* offer unique sources of material. The library staff of a public library has a very distinct set of priorities, and public library holdings will share little or no common ground with the collection of a major university. This is, however, the strength of such a collection. At my local university, for example, the only literature concerning residential electrical wiring or plumbing or framing construction methods is located in one of the campus branch libraries that serves the architecture program. I had much better luck at the public library.

Large or small, every one of the libraries in your area is organized in exactly the same way. They all use one of two highly organized systems for filing books, most likely the Library of Congress system. The LC codes for engineering are identified here, and you can photocopy this list and carry it with you.

The T Classification

T	**Technology (General)**
TA	**Engineering (General), Civil Engineering (General)**
TC	**Hydraulic Engineering**
TD	**Environmental Technology, Sanitary Engineering**
TE	**Highway Engineering. Roads and Pavements**
TF	**Railroad Engineering and Operation**
TG	**Bridge Engineering**
TH	**Building Construction**
TJ	**Mechanical Engineering and Machinery**
TK	**Electrical Engineering, Electronics, Nuclear Engineering**
TL	**Motor Vehicles, Aeronautics, Astronautics**
TN	**Mining Engineering, Metallurgy**
TP	**Chemical Engineering**
TR	**Photography**
TS	**Manufacturing**
TS	**Packaging**

Most of the sciences are in the Q classification; however, medicine has its own category—R, and psychology is in the B classification. A few other scientific disciplines appear elsewhere. Oceanography, anthropology, and physical geography are in the G classification, for example.

The Q Classification

Q	Science (General)
QA	Mathematics
QB	Astronomy
QC	Physics
QD	Chemistry
QE	Geology
QH	Natural History
QK	Botany
QL	Zoology
QM	Human Anatomy
QP	Physiology
QR	Bacteriology
R	Medicine (General)
RC	Internal Medicine
RM	Pharmacology

The two lists of codes are the key to the library. Do not hesitate to ask a librarian for help. There is usually an information desk right beside the indexes or electronic resources you will want to examine. Large libraries can be daunting, but notice that everyone is off in a corner somewhere. That is because there is only one corner of the library that is of concern to each student. Once you find out where your interests are shelved, that is all there is to it. If you are feeling a little overwhelmed, head for the periodicals room and look up all the magazines in your field. This task is easy as well as entertaining, *and* it is very useful because the most timely research—in computer technology, for example—will be found only in magazines. Magazines are important resources.

In technical fields, it may be possible to gather information from a *company library* where you work or from government agencies. Many corporations maintain libraries. You can assume that access to such a resource is usually restricted, but some state universities keep track of the locations of corporate libraries nonetheless. I have a list of all the corporate libraries in the state of Washington. The list was available at the University of Washington in Seattle. The value of such a list is that it usually will include state and national *agency libraries* that may be in your area. These agencies are more likely to maintain libraries of materials and files of documents that are publicly available. For example, if I were interested in oceanography or climatology, I would certainly take the time to check on the large NOAA campus here in Seattle. The National Oceanographic and Atmospheric Administration is an important federal agency serving these sciences.

Another source of material is the *publications of agencies,* both at the state and federal levels. Federal agencies conduct extensive publishing programs, and the cost of their publications is often quite modest. NASA, for example, has published beautiful books of photographs from their planetary missions, and the agency publishes a magazine for engineers. Many agencies at the local level may be of interest to civil engineers, mechanical engineers, and other engineers who frequently interact with these organizations in one way or another.

If a corporate library is closed to the public, the same company might gladly provide you with pounds of *company catalogs* and *brochures,* as well as parts lists, maintenance manuals, mechanical or electromechanical drawings, schematics, and an endless stream of materials that support whatever the company manufactures. As you become more acquainted with your area of interest, you will recognize those companies that are the major players among the industries that are part of your technical specialty. In electronics, it might be Hewlett Packard. In HVAC, it might be Lennox. In chemistry, it might be Sigma-Aldrich. In computers it might be Apple or IBM. You will find that the companies are very supportive as long as the information you request is not proprietary. Bicycle manufacturers, for example, will send you lots of material—as long as you do not ask for the composition of some exotic alloy the company is using on, let's say, a bicycle frame. You probably could not request construction specifics for patented suspension forks. A number of preengineering students in my classes—usually students who are interested in mechanical engineering—have run into dead ends because of levels of information that are proprietary, such as both the bicycle examples.

Trade Journals and Society Journals

Even though there are probably millions of books, magazines, and other documents within a few miles of where you are sitting, libraries cannot encompass every imaginable technical field. For example, if you are interested in material in Fire Command and Administration, there are at least four national magazines related to the field, but neither a university library nor the city library will have them. These magazines are what are called *trade journals.* Here are a few of the hundreds of these important magazines:

Fiberoptic Product News	*NASA Tech Briefs*
C&EN/Chemical Engineering News	*Semiconductor International*
Environmental Protection	*Microwave Journal*
Process Heating	*Military & Aerospace Electronics*
Pollution Engineering	*Solid State Technology*

Modern Plastics	*Digital News & Review*
Laser Focus World	*Electronic Servicing and Technology*
Software Magazine News	*Netware Solutions*
Compliance Engineering	*Chemical Engineering*
Machine Design	*Personal Engineering and Instrumentation*
LAN Times	*Biomedical Products*
Communication Systems Design	*Robotics World*

Another source of information is the *professional engineering society* that serves your special area of interest. Most of the engineering societies are national ones, with local branches in major cities. Of significance to you is that the societies publish magazines and newsletters (usually both) that are the highlight of membership. Not only do the magazines focus on specific areas of interest, they also list job openings, news about projects, and stories about members. For example, if you joined the Professional Electronics Technicians Association, you would receive a magazine every two months and a periodic newsletter as well. You will also find that many of the societies offer student memberships at *very* substantial savings. You are under no obligation to attend meetings.

To give you an idea of the vast number of professional engineering societies, here is a list of those that meet monthly in my area:

AACE: American Association of Cost Engineers	**IEEE: Institute of Electrical & Electronics Engineers**
ACS: American Ceramic Society	**IESNA: Illuminating Engineering Society of North America**
ACSM: American Congress on Surveying & Mapping	**IIE: Institute of Industrial Engineers**
AFS: American Foundrymen's Society	**ISA: Instrument Society of America**
AIAA: American Institute of Aeronautics & Astronautics	**ITE: Institute of Transportation Engineers**
AICHE: American Institute of Chemical Engineers	**NACE: National Association of Corrosion Engineers**
AIME: American Institute of Mining, Metallurgical & Petroleum Engineers	**NAPE: National Association of Power Engineers**
AIPE: American Institute of Plant Engineers	**RI: Robotics International of SME**
ANS: American Nuclear Society	**SAE: Society of Automotive Engineers**
APICS: American Production & Inventory Control Society	**SAME: Society of Military Engineers**
	ASCE: American Society of Civil Engineers

ASEE: American Society for Engineering Education

ASHRAE: American Society of Heating, Refrigeration & Air Conditioning Engineers

ASM: American Society for Metals

ASME: American Society of Mechanical Engineers

ASQC: American Society for Quality Control

ASSE: American Society of Safety Engineers

AWS: American Welding Society

AWWA: American Water Works Association

CASA: Computer & Automated Systems Association of SME

ECS: Electrochemical Society

EM: Electronics Manufacturing Association of SME

SAMPE: Society for the Advancement of Material & Process Engineering

SCE: Society of Chinese Engineers

SFTE: Society of Flight Test Engineers

SHPE: Society of Hispanic Professional Engineers

SOLE: Society of Logistics Engineers

SME: Society of Manufacturing Engineers

SPE: Society of Plastics Engineers

SSS: System Safety Society

SWE: Society of Women Engineers

TAPPI: Technical Association of The Pulp & Paper Industry

If you are interested in a highly specialized area of engineering—perhaps ceramics engineering—the trade journals and the society magazines will be your best source of professional information. Regardless of the engineering field of interest, there is usually a society that can be of help. The largest of them, such as IEEE (Institute of Electrical and Electronics Engineering) are indispensable resources because of their large selection of magazines and newsletters. Within the IEEE are *dozens* of smaller societies that have their own journals. Listed next are a few of the subgroups in the IEEE and the title of the magazine each group publishes.

Society	Magazine
Aerospace and Electronic Systems Society	*Aerospace and Electronic Systems Magazine*
Communications Society	*Communication Magazine*
Computer Society	*Computer Magazine*
Control Systems Society	*Control Systems Magazine*

Dielectrics and Electrical Insulation Society	*Electrical Insulation Magazine*
Engineering Management Society	*Engineering Management Review*
Engineering in Medicine and Biology Society	*Engineering in Medicine and Biology Magazine*
Geoscience and Remote Sensing Society	*Transactions on Geoscience and Remote Sensing*
Industrial Electronics Society	*Transactions on Industrial Electronics*
Industry Applications Society	*Transactions on Industry Applications*
Information Theory Society	*Transactions on Information Theory*
Instrumentation and Measurement Society	*Transactions on Instrumentation and Measurement*
Lasers and Electro-Optics Society	*Photonic Technology Letters*
Oceanic Engineering Society	*Journal of Oceanic Engineering*
Power Engineering Society	*Power Engineering Review*

Exercises Chapter 9, Section F

Write a memo to your instructor identifying resources in your area of technical specialization. List resources in your local area, but also include the addresses of societies or associations, and identify relevant publications these organizations make available.

For purposes of this project, see what resources are available from the following sources:

1) Campus library holdings

   (Identify general LC cite numbers)

2) City library holdings (and LC numbers)

3) Other college and university libraries holdings (and LC numbers)

4) Campus library periodicals

5) City library periodicals

6) Other institutions' periodicals

7) Society and association publications.

Briefly describe the types of materials you find at each resource and explain where the material is located (within the facility).

This project should be no more than 500 words, typed and double-spaced. It will consist of one or two pages.

Documentation
in Detail

Depending on your special area of interest in engineering, you may want to consult style manuals that explore documentation in greater depth. One or more of the following volumes may be of interest to you.

Biology

Council of Biology Editors. *Scientific Style and Format: The CBE Manual for Authors, Editors, and Publishers.* **6th ed. Edited by Edward J. Huth. Bethesda, MD: Council of Biology Eds., 1994.**

Chemistry

American Chemical Society. *ACS Style Guide: A Manual for Authors and Editors.* **2nd ed. Edited by Janet S. Dodd. Washington, DC: American Chemical Society, 1997.**

Electronics

Information for IEEE Transactions, Journals and Letters Authors. **Piscataway, NJ: Institute of Electrical and Electronics Engineers, 1998.**

Geology

U.S. Geological Survey. *Suggestions to Authors of the Reports of the United States Geological Survey.* **7th ed. Washington: GPO, 1991.**

Mathematics

American Mathematical Society. *AMS Author Handbook.* **Providence, RI: American Mathematical Society, 1997.**

Medicine

American Medical Association. *Manual of Style: A Guide for Authors and Editors.* **9th ed. Edited by Cheryl Iverson. Chicago: American Medical Association, 1997.**

Physics

American Institute of Physics, Publications Board. *Style Manual for Guidance in the Preparation of Papers.* **4th ed. New York: American Institute of Physics, 1990.**

Psychology

American Psychological Association. *Publication Manual of the American Psychological Association.* **4th ed. Washington, DC: American Psychological Association, 1994.**

A popular general guide to footnote and bibliography practices is the MLA guide, which is used by most English instructors who teach research-writing courses.

MLA Handbook for Writers of Research Papers. **5th ed. New York: Modern Languages Association, 1999.**

For up-to-date instructions on citing electronic sources, consult the MLA Web site at

http://www.MLA.org/main_stl-nf.html

Guides to using electronic sources include *Online: A Reference Guide to Using Internet Sources* by Andrew Harnack and Eugene Kleppinger (New York: St. Martin's Press, 1997). This book includes guidelines for documenting electronic resources from a number of publishing organizations.

Internet Navigation with Search Engines

If you have not already signed up, consider opening a Telnet account through your campus computing services, or pay the modest computer lab fee on campus. If you are fortunate the lab fee will include access to the Internet through one of the popular browsers such as Netscape or Microsoft's Explorer. Netscape is used most often on campus, and it is a desktop icon in the campus labs where I work. You can also access the Web from home if you have an ISP account (an Internet Service Provider).

The Internet has given computers a totally different utility in recent years. Several generations of computer users were brought up to think of the computer as a processor of input—*their* input. Even if users use the computer only as a word processor, as in my case, they still think of the computer as a tool that serves the user, who acts as a resource. The Internet has reversed this image. The minute you click on the browser, everything is *output*. *You* do not do a thing. A browser makes your computer a resource—and an enormous one at that. A browser provides a version of channel hopping on your TV, except that television has no biomed shows, HVAC shows, chem shows, or EE shows. But there is plenty of technical fare on the World Wide Web. The Web can be pure business, and you will find biomed, HVAC, chemistry, and electronic sites along with hundreds of other engineering, scientific, and technological areas of interest.

Earlier, I referred to library research as legwork. Using a Web browser can be a rewarding approach to research and one that takes less commuting time. But time it does take. Once you key in a URL, the browser is oh so fast, but *finding* the URL in the first place can be oh so slow. Since your research interests are technical, there are shortcuts you can use to avoid search engine results that provide you with a thousand site listings.

Web Sites

Because corporate literature, trade journals, and professional societies will be significant research points of contact in your research, you should locate the Web sites of the key players in your field and enter them as bookmarks so that you can call up the sources whenever you want them. (The telephone, oddly enough, may be the fastest way to find the Web site number for a corporation.) For example, if you have an interest in computer technology, the following listings might be a few of your bookmarks.

Gateway 2000 Inc.	**http://www.gateway.com**
Hitachi PC Corporation	**http://www.hitachipc.com**
Iomega Corp.	**http://www.iomega.com**
Syquest Technology Inc.	**http://www.syquest. com**
Micron Electronics, Inc.	**http://www.micronpc.com**

Trade magazines and society magazines may list their Web sites on the table of contents page or wherever you see the statement of ownership—the fine print that lists the company address and phone numbers. The computer-related bookmarks could include trade magazines such as the following:

Mac World	**http://macworld.zdNet.com**
PCWorld Online	**http://pcworld.com**

As an example of a professional society site, here is the location for the publications of the American Mathematical Society:

American Mathematical Society   **http://www.ams.org**

The home page of each important resource is all I usually enter as a bookmark. Here is the URL for the Army Corps of Engineers:

Army Corps of Engineers   **http://www.hnd.usace.army.mil**

Be sure also to ask your instructors for Web sites that would be of interest in your engineering area. If an instructor has a computer on his or her desk, there is a good chance that there are many useful bookmarks on the browser.

Directories and Engines

Other useful Web sites are those that list databases of information in your area of interest. A half dozen such sites might greatly abbreviate your research needs. Begin by consulting a Net directory such as the World Wide Web Virtual Library or Yahoo! The Virtual Library is the oldest catalog on the Web, and it was started by Tim Berners-Lee, founder of the Web itself.

Virtual Library   **http://www.vlib.org**

Yahoo!   **http://www.yahoo.com**

Using the Virtual Library you can find a number of science and engineering sites that are focused on specific fields. For example, you might be interested in the following sites:

Astronomy	http://Webhead.com/WWWVL/Astronomy/astro.html
Biology	http://mcb.harvard.edu/biolinks.html
Biosciences	http://www.vlib.org/Biosciences.html
Chemistry	http://www.chem.ucla.edu/chempointers.html
Computer Science	http://www.vlib.org/Computing.html
Engineering: General	http://www.vlib.org/Engineering.html
Civil & Environmental	http://www.ce.gatech.edu/WWW-CE/home.html
Chemical	http://www.che.ufl.edu/WWW-CHE
Electrical	http://webdiee.com.itesm.mx/diee/wwwvlee
Mechanical	http://CDR.stanford.edu/WWW-ME/home.html

Software	http://rbse.jsc.nasa.gov/virt-lib/soft-eng.html
Engineering Technology	http://arioch.gsfc.nasa.gov/wwwvl/engineering.html
Mathematics	http://euclid.math.fsu.edu/Science/math.html
Mechanical Engineering Societies	http://cdr.stanford.edu/WWW-ME/special.html
Physics	http://www.fisk.edu/vl/Physics/Overview.html

The Yahoo home page is illustrated in Figure 10-1.

Figure 10–1   The Yahoo! Home Page (Reproduced with permission of Yahoo! Inc. © 1999 by Yahoo! Inc. YAHOO! and the YAHOO! logo are trademarks of Yahoo! Inc.)

The well-known disorganization that characterizes the Web is the sort of chaos that can absorb a lot of research time. By using highly organized resources such as the Net directories, you can restrict wide-open searches that waste time. To use the Internet as a library, you need to find highly orderly indexes to sources in order to search precise databases. Large subject indexes, such as the Virtual Library and the specialized indexes of professional societies and organizations, will probably serve to organize your efforts. The Virtual Library index for engineering is illustrated in Figure 10-2.

The most important Web directories are a good place to start research. Here is a list of the large directories that are frequently consulted. The Web home page is identified on the left. Science and Technology sectors are indicated on the right.

The Virtual Library:
Engineering

<u>Engineering</u>

- <u>Acoustics and Vibrations</u>
- <u>Aerospace</u>
- <u>Architecture</u>
  - o <u>Architectural Engineering</u>
  - o <u>Landscape Architecture</u>
- <u>Cartography</u>
- <u>Ceramics</u>
- <u>Chemical Engineering</u>
- <u>Civil Engineering</u>
- <u>Control Engineering</u>
- <u>Cross-Connection Control/Backflow Prevention</u>
- <u>Electrical Engineering</u>
- <u>Engineering and Technology Management</u>
- <u>Geotechnical Engineering</u>
- <u>Hazards and Risk</u>
- <u>Industrial Engineering</u>
- <u>Manufacturing Engineering</u>
- <u>Materials Engineering</u>
- <u>Mechanical Engineering</u>
- <u>Naval Architecture & Ocean Engineering</u>
- <u>Nuclear Engineering</u>
- <u>Optical Science and Engineering</u>
- <u>Petroleum and Geosystems Engineering</u>
- <u>Software Engineering</u>
- <u>Standards and Standardization Bodies</u>
- <u>Technology Transfer</u>
- <u>Telecommunications</u>
  - o <u>Amateur Radio</u>
- <u>Wastewater Engineering</u>
- <u>Welding Engineering</u>

Figure 10-2 The Virtual Library Engineering Index. (Illustration Courtesy of the WWW Virtual Library.)

Home Page	Engine	Engineering/Science Directories
http://www.yahoo.com	Yahoo!	http://dir.yahoo.com/Science/Engineering/
http://www.lycos.com/	Lycos	http://www.lycos.com/wguide/network/net_484378.html http://dir.lycos.com/Science/
http://galaxy.com/	Galaxy	http://galaxy.com/galaxy/Engineering-and-Technology.html http://galaxy.com/galaxy/Science.html
http://www.altavista.com	AltaVista	http://www.altavista.com/cgi-bin/query?pg=q&kl=XX&q=engineering http://www.altavista.com/cgi-bin/query?pg=q&kl=XX&q=science
http://www.hotbot.com	Hotbot	http://directory.hotbot.com/Science/
http://www.infoseek.com	Infoseek	http://infoseek.go.com/WebDir/Science/Engineering
http://www.studyweb.com	StudyWEB	http://www.studyweb.com/science/

These engines are very powerful, by which I mean they can search not just millions of sources, but tens of millions.

> **Note:** Many Web addresses found in this chapter were originally provided with links that had to be removed because of frequent changes. You will have to search under the domain name to see where the server has placed the information you seek.

Yahoo!

Originating at Stanford University, in Palo Alto, California, Yahoo is widely viewed as the biggest search engine (as of this writing*). The system will search its enormous database categories using Boolean operators (*and, or, not*) to help you trim the search (see p. 320).

Lycos

Another one of the largest search systems, Lycos was developed by Carnegie Mellon University in Pittsburg, Pennsylvania. Lycos Inc. is located there. The searches will handle Boolean queries.

Galaxy

Galaxy has a fairly substantial index. The Galaxy directory to engineering and technology is illustrated in Figure 10-3.

* All data presented here are subject to rapid changes.

galaxy *The professional's guide to a world of information.*

Home
What's New
Add Your Site
Advanced Search
Info & Help

 *One of These Children is Autistic...*

Engineering and Technology

- **Agricultural Engineering**
- **Bioengineering**
  Biomedical Engineering
- **Chemical Engineering**
  Biotechnology
- **Civil and Construction Engineering**
  Architectural Engineering - Energy Science - Environmental Science - Nuclear Engineering - Petroleum Engineering
- **Computer Technology**
  Adaptive Systems and Artificial Life - Artificial Intelligence - Computer Aided Design and Engineering - Computer Graphics and Art - Computer Mediated Communication - Control Systems - Database Systems - Formal Methods - High Performance Computing - Human - Computer Interaction - Image Processing - Information Retrieval and Categorization - Multimedia - Music - Networking and Telecommunications - Object Oriented Systems - Programming Languages - Publishing Systems - Real time Systems - Robotics - Security - Simulation and Modelling - Software Archives - Software Engineering - Standards - Storage Systems - User Groups - Virtual Reality - Visualization and Imaging

- **Design Methodologies**
- **Electrical Engineering**
  Compliance Engineering - Electronics Engineering
- **Human Factors and Human Ecology**
  Ergonomics - Industrial Design
- **Manufacturing and Processing**
  Corrosion Engineering - Cryogenics - Industrial Engineering - Metallurgical Engineering - Paints and Coatings - Quality Control
- **Materials Science**
- **Mechanical Engineering**
  Computational Mechanics - Materials Engineering - Nanotechnology - Robotics
- **Nondestructive Testing**
- **Technical Reports**
- **Technology Transfer**
- **Transportation**
  Aerospace and Aeronautical Engineering - Automotive Engineering

Figure 10-3 The Galaxy Index Page for Engineering. (Illustration Courtesy of Galaxy. Copyright © 1993–1999 AHN Partners, L.P. All rights reserved.)

AltaVista

Alta Vista emerged in 1995 at Digital Research Laboratories in Palo Alto, California. Its Web index evidently now includes in the area of 140 million Web pages and has far exceeded the 11 billion words indexed on the 30-gigabyte Web index system that it used several years ago. AltaVista also contains an index to over 14,000 news groups and can produce the full text of any articles it finds. This engine is considered strong in scientific information.

HotBot

HotBot indexed over 110 million Web pages and "refreshes" the entire database every month. The site is intent on full coverage of Usenet Groups. It is owned by Lycos Inc. The search engines accept Boolean operators, which greatly aid a search. Alta Vista and HotBot offer a third of the Web's total material.

Infoseek

Infoseek was founded in 1994 and is located in Santa Clara, California. The corporation provides 50 million indexed documents and 80 million URL addresses. The collection includes the Web, Usenet news groups, company files, and other varieties of Internet resources.

There are other popular directories, and some do not include engineering and science categories in their home page listings. Type the keyword *engineering* or *science* or *education* in the search window to check the basic availability of material.

World Wide WebWorm	**http://goto.com**
LookSmart	**http://www.looksmart.com**
Argus Clearing House	**http://www.clearinghouse.Net**

Metaengines

The metaengines have emerged in the last few years; they enter other engines and compile data from them. These systems—called "parallel Internet query engines"—are becoming numerous.

The metaengines listed here will appear similar, but each one has unique search features and other conveniences. SavvySearch, for example is available in languages other than English. It also searches the DejaNews index of Usenet sites. Profusion will send you updates. Some of the systems "score" the search results, particularly if they list material from other engines that rank the importance of their search findings. Such engines rank the materials by "relevance," but only insofar as the engines locate the frequency of keywords, the number of visits or "hits," or other statistical measures. All of them have their own individualized search approaches. There are, in fact, several technologies behind the engines that make each one unique.

The Internet Sleuth	**http://www.the bighub.com**

The Sleuth search engine will query through a number of other engines including AltaVista, Excite, Infoseek, HotBot, OpenText, Lycos, and WebCrawler.

MetaCrawler	**http://www.metacrawler.com/**

This metaengine emerged in 1995 from the University of Washington in Seattle. It is now owned by Go2Net Inc. The multiple-engine search (nine of them) will access many of the major engines including AltaVista, Infoseek, Lycos, Yahoo!, Excite, WebCrawler, and Inktomi (a search engine from the University of California at Berkeley).

SavvySearch   http://www.savvysearch.com

Another multiple engine search system, SavvySearch was designed by a former student from Colorado State University. It will gather responses to a query from the following sources: AltaVista, Excite, Lycos, WebCrawler, Infoseek, HotBot, and OpenText. It has the additional search capability to seek out Usenet resources from DejaNews (a search engine that handles Usenet locations).

If you decide to go straight to the search engine services, you should trim the search by using the same hierarchical system that is used by the directories, which will focus the subject area. Since more than a few hours are needed to become familiar with the directories and the search services, experiment with one of the metaengines that tap many of the other directories and engines. You might try EZ-Find, still another one of the metaengines. Go to the EZ-Find keyword entry form and create a word string of interest in your field. You can go through a series of search buttons to see how the different services handle your specialty. Some services might be very weak in your subject area; others might be very strong. EZ-Find and the other metaengines are ways to establish some familiarity with the various services. Go to the services directly once you see who shares your interest.

EZ-Find   http://info.theriver.com/TheRiver/ezfind.htm

Using The Net

Surfing! Sunshine, fresh air, not a care in the world, and plenty of time on your hands. This probably is not an accurate description of an engineering tech student—especially the idea that a college student has plenty of time to surf, either the waves or the Internet. It is for this reason that most of the preceding comments are intended to focus your time and energy into four main *domains*.

.com	commercial
.edu	educational
.gov	government
.org	nonprofit organization

You will find many professional organizations of interest in the .org category. The following sites are but a few of the many engineering-related organizations that were identified earlier in Chapter 9. The home page of the American Institute of Chemical Engineers is shown in Figure 10-4.

American Ceramic Society	**http://www.acers.org/**
American Institute of Chemical Engineers	**http://www.aiche.org/**
American Society for Engineering Education	**http://asee.org/**
American Society of Mechanical Engineers International	**http://www.asme.org/**
American Society of Civil Engineers	**http://www.asce.org/**
American Association for the Advancement of Science	**http://www.AAAS.org/**

There are two reasons to limit your research in other areas of the Net. Time is one; *Usenet news groups* and *listserve mailing lists* may be slow avenues when you realize that 10 *million* computers are linked to the Web. The other problem is professionalism. Your needs are technical and specific, and you want the best resources available. The Internet has no quality controls built into it except for some specific groups that are *policed* or restricted lists that are *moderated*. There are thousands of newsgroups, and the numbers indicate their popularity. In the *alt* category alone, there are several thousand. There is a science category of newsgroups, of course, and you might see what is of interest in the *sci* category.

If you are new to the Web, a few suggestions might be useful. Getting *to* your resources is one issue. Getting *into* and *out* of the source is another. *Hypertext* is the unique screen display that is used on the Net. It is an electronic page of sorts, but it has fast-forward features built into keywords in the text and icon cues that surround the page. If your cursor turns to a hand, its index finger is ready to press the word or icon and fast-forward to another document or section of a Web site. The invention of hypertext was a mixed blessing. Opening pages are cluttered with the icons because these links are basically an index to other sections of text. A problem that can occur as you click through successive links is that you get lost because the process is not linear. The Netscape toolbar, shown in Figure 10-5, will rescue you.

In the upper left corner, the Back icon will exactly reverse the path you have taken to a page, and it will show each page location in the location window. The history menu on Internet Explorer, shown in Figure 10-6, also keeps a list of any sites you visited.

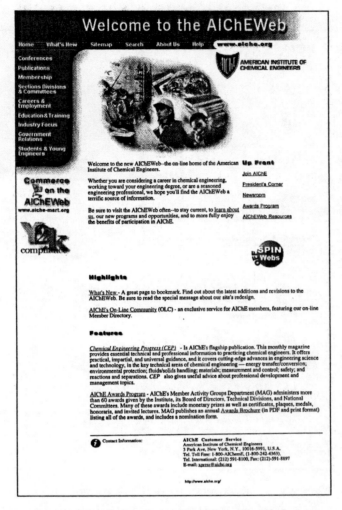

Figure 10-4 Home Page for the American Institute of Chemical Engineers.
(Illustration Courtesy of AIChE.)

Here is a typical location clicked deep in the maze of a Web site:

http://netbuyer.zdnet.com/texis/netbuyer/ ddoframe.html?u=www.zdnet.com/netbuyer/edit/ howtobuy/index.html&rg=r129

You can bookmark a specific site when desired. Remember to do so! The preceding Web site was not saved as a bookmark, and there is an error somewhere in the listing. The number is now lost forever; it is hoped that computer technology will streamline these site location numbers in the future. In the meantime, you can simplify the site numbers by avoiding *links* as your site reference.

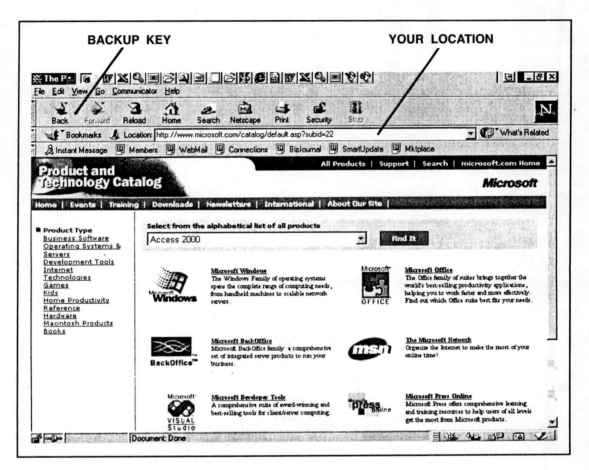

Figure 10-5 Netscape Controls. (Screen shot reprinted by permission from Microsoft Corporation.)

Remember that whenever you enter a query you are in a database, and the search is conducted only in that database. The more specific your database, the easier it is to use the search feature. If you have three weeks to write a paper, you do not want a list of 500 prospects to show up on the screen. Ten or so precise items would be ideal.

If keyword searching is new to you, an Internet guide will help you streamline your efforts. Here are three basic tips.

- Use multiple word searches (strings) rather than simple words.

- Use the word *and* between the words so that the search calls up only those resources that use the full string of words.

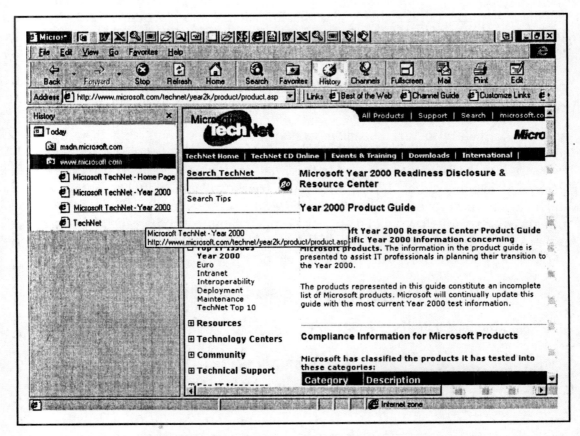

Figure 10-6 Internet Explorer with the History Menu Opened on the Left. (Screen shot reprinted by permission from Microsoft Corporation.)

- Place the string in "quotation marks" so that the resources are called up that use the exact phrase you quoted.

Web research is exciting, and the search engines are astonishing tools. If you have a paper due in a few weeks, you want to shave off as many hours as you can in pursuit of material. The charm of search engines can be deceiving. Focus on the professional societies and companies of interest and Web directories to areas in your field. First, consult important bookmarks that you have saved. Use the search engines second. Narrow the focus; otherwise the engines may list so many sites that they will negate the advantage of using the speed of the Net.

If you are new to the Web try this site for a starter. Open your browser and enter this URL:

http://www.scorecard.org

You will find directions for locating the worst pollution health hazards in the nation, including those near you. The chemical scorecard will use your zipcode to show you a map that includes *your* street location and identifies any nearby risk sites. This service of the Environmental Defense Fund is contacted thousands of times every day. I hope you are not downwind of a hazardous site.

Internet Updates

All the information in the preceding section and the upcoming section is subject to change. The Internet is a free-market technology and remains unregulated and very much in transition. If you experience an occasional frustration over site deletions or site changes, delete the address links after *.org* (or *.gov, .com,* and so on) on your site address and try again. This procedure shortens an address to the domain name (or the home page). In other words, hyperlink addresses change much more often than the domain names, and the domain names will usually allow you to locate what you are after. For example, an older site address for the Acoustical Society of America

http://asa.aip.org/standards.html

no longer functions, but the domain name remains the address for the society.

http://asa.aip.org/

If your search yields the following message

NOT FOUND

The requested URL/.std html was not found on this server.

---

you know the domain is active because the server of the domain has been consulted for the search.

Other sites that may be of value are the indexes to search engines and updates concerning changes among them.

- For an index of over 1000 engines classified into 50 categories try:

Virtual Newest Search Engine   **http//www.dreamscape.com/ frankvad/search.newest.html**

- A large list of the engines is also maintained at the following Yahoo! site:

http://dir.yahoo.com/Computers_and_Internet/Internet/
World_Wide_Web/Searching_the_Web/All_in_one_Search_Pages/

- Perhaps the most important site concerning the engines is Search Engine Watch. This site has a monthly newsletter also.

Search Engine Watch   **http://searchenginewatch.com**

Exercises Chapter 10, Section A

Complete one of the following exercises:

---

1. Compare two of the well-known engines identified in this chapter. Feel free to use other engines with which you are familiar. Copy the engineering index from each engine if one has been provided, and then copy the lead pages of the link that is most appropriate for your engineering or technical discipline. Browse for a few points of interest and see how the engines differ in services and results. Explain your findings in a memo to your instructor and include the screen scans as "attachments."

---

2. This instructional activity may be of value to you and your classmates. If some members of the class are unfamiliar with the Internet, teams of two can go to the computer lab and explore the Web to locate some of the basic sites listed in this chapter. Each team consists of a "novice" and someone familiar with the Internet. The object is for one to explain to the other how to browse the Web. The novice can take notes and write a brief memo to the instructor explaining what procedures he or she learned. The student who helps guide the novice has no assignment other than the role of volunteer teacher.

Finding Engineering Sites on the Web

The following pages contain a number of lists that should help you get started in your efforts to develop an efficient, Internet library of dependable references. You will find the Web resources useful for both your college work and your professional work. Undoubtedly some of the resources will be as interesting as favorite magazines, and you will pull up the sites at work or at home to read about the latest issues in your engineering field.

The information is organized more or less in the order covered in the last chapter. Internet resources may be electronic, but they correspond to the basic traditional avenues of research:

> **Libraries**
>
> **Trade Magazines and Society Journals**
>
> **Professional Associations**
>
> **Engineering Standards**
>
> **Usenet Groups**
>
> **Listservs**
>
> **Research Databases**

Libraries

University libraries with Web sites can be of particular value for compiling a bibliography of resources if they have the on-line technical facilities for file searches and if there are no restrictions to searching their catalogs. Several WWW library addresses are specific to engineering. As I mentioned earlier, access to actual documents is probably not possible for anything that is not available on-line.

Edinburgh Engineering Virtual Library
http://www.eevl.ac.uk/

Internet Connections for Engineering (compiled at Cornell University)
http://www.englib.cornell.edu/ice/

Penn Library Engineering Page
http://www.library.upenn.edu

Cyberspace libraries with engineering specializations are based at a number of locations, many of which are universities. Sites for specific engineering fields that are of interest include the following. There are, of course, dozens of engineering specializations, and only the major fields are represented here. Figure 10-7 shows the home page of the Mechanical Engineering Virtual Library.

Acoustics and Vibration Virtual Library
http://www.ecgcorp.com/velav/

Aeronautics and Aeronautical Engineering Virtual Library
http:/macwww.db.erau.edu/www_virtual_lib/aeronautics.html

Aerospace Engineering Virtual Library
http://macwww.db.erau.edu/www_virtual_lib/aerospace.html

Architecture, Landscape Architecture Virtual Library
http://www.clr.toronto.edu:1080/VIRTUALLIB/archGALAXY.html

Chemical Engineering Virtual Library
http//www.che.ufl.edu/WWW-CHE/index.html

Civil Engineering Virtual Library
http://www.ce.gatech.edu/WWW-CE/home.html

Control Engineeering Virtual Library
http://www-control.eng.cam.ac.uk/extras/Virtual_Library/
Control_VL.html

Geotechnical Engineering Virtual Library
http://geotech.civen.okstate.edu/wwwVL/index.html

WWW Virtual Library | Home | Free Online | University ME | ME Institutes | ME
Mechanical Engineering | | Services | Departments | & Societies | Vendors

World Wide Web Virtual Library
Mechanical Engineering

Search the WWW Virtual Library

Match: [All] [⬍] Format: [Long] [⬍]

Database: [The WWW VL for Mechanical   Engineering] [⬍]

Search: [_____] [Search]

University ME Departments   **Free On-Line Services**
ME Institutes and Societies   • General
ME Vendor Pages   • Jobs
  • Mechanical Components   • Publications
  • Electronic andElectromechanical   • Online Calulations
  • Software   • Catalogs and Databases
  • Educational   • Free Software and Information Services
  • Consulting   • Misc. Information and Services
  • Publications   **Broader Topics**
ME Forums and Newsgroups   **Information / Help**

[∧ Top ∧]

[Index | Online Services | University Depts. | Institutes & Societies | Forums & Newsgroups |
| Vendors | Broader Topics | Contributions | Info / Help]
Administrator: Charles Petrie (Biography) - Webmaster: webmaster@vlme.com
© Copyright 1999 by VLME LLC - Legal Notice - Design by WebSiteMechanic

Figure 10-7 Mechanical Engineering Site of the Virtual Library. (Illustration Courtesy of VLME LLC.)

Mechanical Engineering Virtual Library

http://vlme.org

Optical Engineering Virtual Library

http://www.spie.org/wwwvl_optics.html

Power Engineering Virtual Library

http://www.analysys.com/vlib

Rapid Prototyping Virtual Library

http://arioch.gsfc.nasa.gov/wwwvl/engineering.html#proto

Safety-Critical Systems Virtual Library

http://www.comlab.ox.ac.uk/archive/safety.html

Systems and Control Engineering Virtual Library

http://src.doc.ic.ac.uk/bySubject/Computing/Overview.html

Trade Magazines and Professional Journals

For periodical literature and professional journals, consult the society in your area of specialization or consult one of the following periodical indexes. Figure 10-8 shows the home page of the University of Saskatchewan electronic journal index.

Academic and Reviewed Journals from Virtual Library

http://www.edoc.com

Electronic Journals, Magazines, and Newspapers

http://library.usask.ca/ejournals/

National Science Foundation

http://www.eng.nsf.gov/links/li00005.htm

The search engine ZDNet will search the contents of popular computer technology magazines. Since this is a rapidly changing technology, this engine is a way to locate timely materials if your interests concern computers, computer applications, and network systems.

ZDNet

http://www5.zdnet.com/findit/search.html

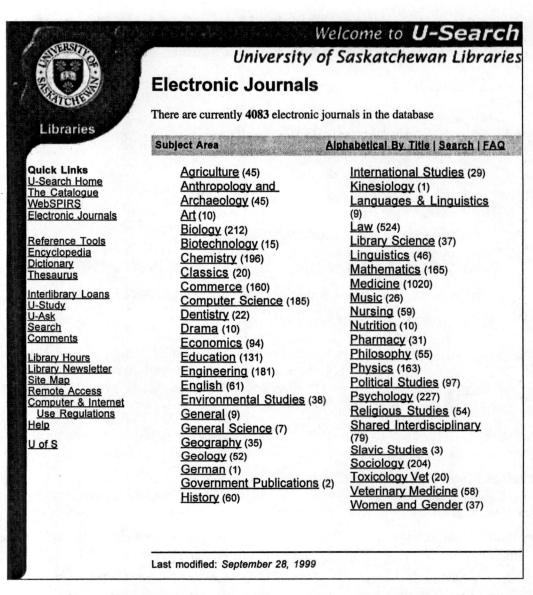

Figure 10-8 Home Page of an Electronic Journals Index. (Illustration Courtesy of the University of Saskatchewan Libraries.)

Professional Associations and Societies

As mentioned earlier, some of the most subject-specific research can be found in the *trade and professional journals* of professional associations and societies. A number of these periodicals are now available from Web sites. A society's home page will also explain other features and opportunities that are available by way of the society's Internet address. For locating the societies of interest to you, consult the following indexes:

Engineering Associates and Institutes
http://www.aecinfo.com/eng/assoc.html

IEEE Societies
http://www.ieee.org:80/society/

Professional Societies: Engineering-Related Sites
http://www.interpath.net

Scholarly Societies
http://www.lib.uwaterloo.ca/society/

Yahoo!: Science: Enginering: Organizations: Professional
http://www.yahoo.com/text/Science/Engineering/
Organizations/Professional/

Among the individual societies, you will find the following addresses. Most of these societies offer journal publications, and there is a trend toward on-line journals and other on-line publications as well. Figures 10-9 through 10-11 show the home pages of three of these professional organizations.

Acoustical Society of America
http://asa.aip.org

American Association for the Advancement of Science
http//www.AAAS.org

American Chemical Society
http://www.acs.org

American Computer Scientists Association Inc. (ACSA)
http://www.acsa2000.net/

American Consulting Engineers Council
http://www.acec.org/

American Electronics Association
http://www.aeanet.org/

American Institute of Aeronautics and Astronautics (AIAA)
http://www.aiaa.org/

American Institute of Architects
http://www.aiaonline.com/

American Institute of Chemical Engineers
http://www.aiche.org/

American Society for Testing and Materials (ASTM)
http://www.astm.org/

American Society of Civil Engineers (ASCE)
http://www.asce.org/

The First Society in Computing

- **Membership**
  Professionals
  Students

- **Publications**

- **Digital Library**

- **Special Interest Groups (SIGS)**

- **Conferences**

- **Education**

- **Chapters**

- **Public Policy**

- **Press Room**

- **Awards**

- **Institutions and Libraries**

- **About ACM**

Association for Computing Machinery

Founded in 1947, ACM is the world's first educational and scientific computing society. Today, our members — over 80,000 computing professionals and students world-wide — and the public turn to ACM for authoritative publications, pioneering conferences, and visionary leadership for the new millennium.

Monday, July 26, 1999

What's New!

New ACM Special Interest Group Officers Elected

Uniform Computer Information Transactions Act (UCITA) Updates Final NCCUSL vote on UCITA July 29, 1999

ACM Position on Software Engineering Licensing

Conferences: [Registration in e-store] [listings in Conference Calendar]

Digital Library Enhancements

Funding Opportunities: [Database] [Email Notification]

Student Contests: [Windows CE Winners] [Programming Contest]

Join
ACM
SIGs

Purchase
Books
Conference
Proceedings

Subscribe to
Digital Library
Journals
Magazines

View
Conference
Calendar

Career
Opportunities

Manage your acm.org account

Contact ACM

ACM - Association for Computing Machinery || Read the ACM Privacy Policy and Code of Ethics

Questions? Comments? Contact webmaster@acm.org
Call: 1.800.342.6626 (USA and Canada) or +212.626.0500 (Global)
Write: ACM, 1515 Broadway, New York, NY, 10036, USA

Figure 10-9 Home Page for the Association for Computing Machinery. (Illustration courtesy of ACM.)

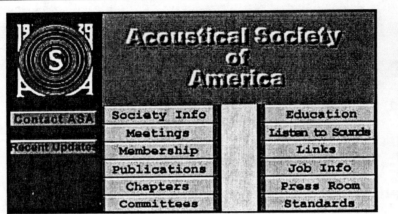

The premier international scientific society in acoustics, dedicated to increasing and diffusing the knowledge of acoustics and its practical applications.

NEW **Columbus Meeting
Information
Registration**

NEW **ARLO
is Here!**

NEW **Go to
JASA Online**

© 1999 The Acoustical Society of America

Figure 10-10 Acoustic Society of America Home Page. (Illustration Courtesy of Acoustical Society of America.)

American Society of Heating, Refrigerating, and Air-Conditioning Engineers, Inc. (ASHRAE)
http://www.ashrae.org/

American Society of Mechanical Engineers (ASME)
http://www.asme.org/

American Society of Professional Engineering (NSPE)
http:www.nspe.org/

American Vacuum Society
http://www.vacuum.org/

American Water Works Association
http://www.awwa.org

Association for Computing Machinery (ACM)
http://www.acm.org/

Association for Systems Management
http://www.4w.com/asm

Audio Engineering Society
http://www.cudenver.edu/
aes/index.html

Electrochemical Society
http://www.electrochem.org/

Electron Devices Society (EDS)
http://www.ece.neu.edu/eds/EDShome.html

Institute of Electrical and Electronics Engineers (IEEE)
http://www.ieee.org:80/

Institution of Electrical Engineers (IEE)
http://www.iee.org.uk/

Instrument Society of America
http://www.isa.org/

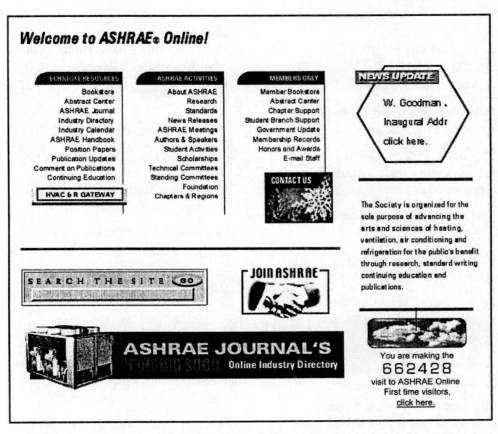

Figure 10-11 American Society of Heating, Refrigerating, and Air-Conditioning Engineers Home Page. (Illustration Courtesy of ASHRAE.)

International Society for Optical Engineering
http://www.spie.org/

International Society for Measurement and Control
http://www.isa.org/

National Association of Corrosion Engineers
http://www.nace.org/

Society of Automotive Engineers
http://www.sae.org:80

Society of Manufacturing Engineers
http://www.sme.org/

World Computer Society
http://computer.org/

Because of its size, the Institute of Electrical and Electronic Engineering has a number of Web sites for various organizations within the parent society.

IEEE Aerospace and Electronic Systems Society
http://aess.gatech.edu

IEEE Antennas and Propagation Society
http://www.ieeeaps.org/

IEEE Broadcast Technology Society
http://www.ieee.org/organizations/society/bt/index.html

IEEE Circuits and Systems Society
http://www.ieee-cas.org/

IEEE Communications Society
http://www.comsoc.org/

IEEE Components Packaging and Manufacturing Technology Society
http://www.cpmt.org/

IEEE Computer Society
http://www.computer.org

IEEE Consumer Electronics Society
http://www.ieee.org/organizations/society/ce/

IEEE Control Systems Society
http://www.odu.edu/~ieeecss

IEEE Dielectrics and Electrical Insulation Society
http://tdei.sju.edu/deis/deishp.html

IEEE Education Society
http://w3.scale.uiuc.edu/ieee_ed_soc/

IEEE Electromagnetic Compatibility Society
http://www.emcs.org./

IEEE Electron Devices Society
http://www.ieee.org/organizations/society/eds/

IEEE Engineering Management Society
http://sils.pratt.edu/ems/index.html

IEEE Engineering in Medicine and Biology Society
http://www.ewh.ieee.org/soc/embs/

IEEE Geoscience & Remote Sensing Society
http://www.ieee.org/organizations/society/grs/index.html

IEEE Industrial Electronics Society
http://www.ewh.ieee.org/soc/ies/

IEEE Industry Applications Society
http://www.ewh.ieee.org/soc/ias/

IEEE Information Theory Society
http://www.itsoc.org/

IEEE Intelligent Transportation Systems Council
http://www.ewh.ieee.org/tc/its/

IEEE Instrumentation and Measurement Society
http://ewh.ieee.org/soc/im/index.html

IEEE Lasers & Electro-Optics Society
http://www.ieee.org/organizations/society/leos/

IEEE Magnetics Society
http://yara.ecn.purdue.edu/~smag

IEEE Microwave Theory and Techniques Society
http://www.mtt.org/

IEEE Nuclear and Plasma Sciences Society
http://hibp7.ecse.rpi.edu/~connor/ieee/npss.html

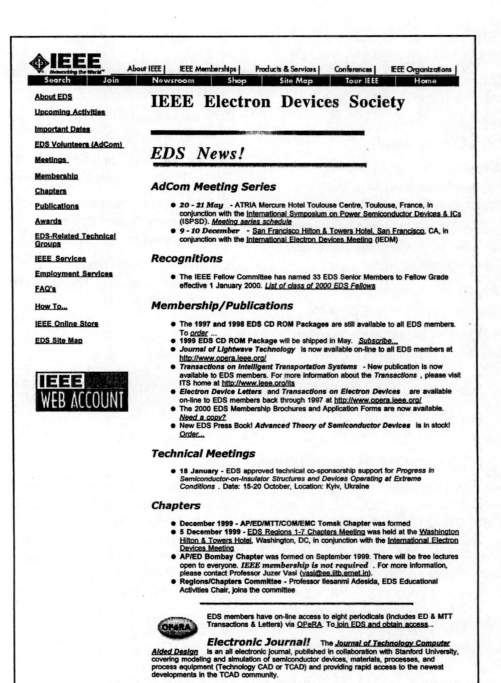

IEEE Electron Devices Society

EDS News!

AdCom Meeting Series

- **20 - 21 May** - ATRIA Mercure Hotel Toulouse Centre, Toulouse, France, in conjunction with the International Symposium on Power Semiconductor Devices & ICs (ISPSD). *Meeting series schedule*
- **9 - 10 December** - San Francisco Hilton & Towers Hotel, San Francisco, CA, in conjunction with the International Electron Devices Meeting (IEDM)

Recognitions

- The IEEE Fellow Committee has named 33 EDS Senior Members to Fellow Grade effective 1 January 2000. *List of class of 2000 EDS Fellows*

Membership/Publications

- The 1997 and 1998 EDS CD ROM Packages are still available to all EDS members. To *order* ...
- 1999 EDS CD ROM Package will be shipped in May. *Subscribe...*
- *Journal of Lightwave Technology* is now available on-line to all EDS members at http://www.opera.ieee.org/
- *Transactions on Intelligent Transportation Systems* - New publication is now available to EDS members. For more information about the *Transactions* , please visit ITS home at http://www.ieee.org/its
- *Electron Device Letters* and *Transactions on Electron Devices* are available on-line to EDS members back through 1997 at http://www.opera.ieee.org/
- The 2000 EDS Membership Brochures and Application Forms are now available. *Need a copy?*
- New EDS Press Book! *Advanced Theory of Semiconductor Devices* is in stock! *Order...*

Technical Meetings

- **18 January** - EDS approved technical co-sponsorship support for *Progress in Semiconductor-on-Insulator Structures and Devices Operating at Extreme Conditions* . Date: 15-20 October, Location: Kyiv, Ukraine

Chapters

- **December 1999** - AP/ED/MTT/COM/EMC Tomsk Chapter was formed
- **5 December 1999** - EDS Regions 1-7 Chapters Meeting was held at the Washington Hilton & Towers Hotel, Washington, DC, in conjunction with the International Electron Devices Meeting
- AP/ED Bombay Chapter was formed on September 1999. There will be free lectures open to everyone. *IEEE membership is not required* . For more information, please contact Professor Juzer Vasi (vasi@ee.iitb.ernet.in).
- **Regions/Chapters Committee** - Professor Ilesanmi Adesida, EDS Educational Activities Chair, joins the committee

EDS members have on-line access to eight periodicals (includes ED & MTT Transactions & Letters) via OPeRA. To join EDS and obtain access...

Electronic Journal! The *Journal of Technology Computer Aided Design* is an all electronic journal, published in collaboration with Stanford University, covering modeling and simulation of semiconductor devices, materials, processes, and process equipment (Technology CAD or TCAD) and providing rapid access to the newest developments in the TCAD community.

Figure 10-12 Home Page for the IEEE Consumer Electronics Society. (Illustration courtesy of IEEE.)

IEEE Neural Networks Council
http://www.ieee.org/nnc/index.html

IEEE Oceanic Engineering Society
http://auv.tamu.edu/oes/

IEEE Power Electronics Society
http://www.pels.org/

IEEE Power Engineering Society
http://www.ieee.org/organizations/
society/power/

IEEE Professional Communication Society
http://www.ieee.org/organizations/
society/pcs/index.html

IEEE Reliability Society
http://www.ewh.ieee.org/soc/rs/

IEEE Robotics & Automation Society
http://www.ncsu.edu/IEEE-RAS/

IEEE Society on Social Implications of Technology
http://www4.ncsu.edu/unity/users/j/
jherkert/index.html

IEEE Solid-State Circuits Society
http://www.sscs.org/info/

IEEE Systems, Man, and Cybernetics Society
http://www.isye.gatech.edu/ieee-smc/

IEEE Ultrasonics, Ferroelectrics, and Frequency Control Society
http://uffc.brl.uiuc.edu/uffc/

IEEE Vehicular Technology Society
http://www.ieee.org/organizations/
society/vts/index.html

Engineering Standards

Many engineering fields abide by various sets of standards or codes. Indeed, the determination of the rigorous standards that codes uphold is an industry in itself for testing engineers and other specialists. Architects and structural engineers could not function without sets of codes, whether they are national electrical codes or local building codes. Several of the organizations involved in such standards are global in reach and set standards for the convenience of international communities in engineering, research, and manufacturing. Figure 10-13 shows the home page of the American National Standards Institute.

American National Standards Institute (ANSI)

http://www.ansi.org/

American Society for Testing and Materials (ASTM)

http://www.astm.org/

American Society of Agricultural Engineers Standards

http://www.asae.org/standards/

Figure 10-13 American National Standards Institute Home Page. (Illustration courtesy of ANSI.)

Codes and Standards of ASME (American Society of Mechanical Engineers)

http://www.asme.org/codes/

IEEE Standards

http://www.standards.ieee.org/index.html

Standards for the National Association of Corrosion Engineers

http://www.nace.org/

Standards of the Acoustical Society of America

http://asa.aip.org/

International Organization for Standardization (ISO)

http://www.iso.ch/

Usenet Groups

The three following Internet resources—Usenet groups, Listserv mailing lists, and research databases—are included here to suggest other approaches to engineering that are available on the Web. Ranging from cyberchat to the most serious and obscure levels of academic research, these sources may prove to be of varying degrees of interest. All these resources exist, of course, because professional engineers create and support these avenues of communications for thousands of unique interests in engineering.

Usenet groups are discussion groups in which people post messages on the Web concerning thousands of topics. Many engineering groups are numbered among them. Engines such as AltaVista and HotBot, for example, will search Usenet sources. There are hundreds of these sites for computer engineering issues alone. In his very useful text *The Internet for Scientists and Engineers,* Brian Thomas created a brief index for many of the sites concerning computer engineering. He used the following categories:

Architectures	**Education**
Compilers	**Emulators**
Compression	**Human Factors**
Databases	**Information Systems**
Data Communications	**Language**
Documentation	**Organizations**

Parallel Processors	**Software Engineering**
Patents	**Sources**
Programming	**Specifications**
Protocols	**Standards**
Real-time Processing	**Systems**
Research in Japan	**Theory**
Risks	**Unix**
Robotics	**Viruses**
Simulation	

There are dozens of newsgroups in some of these computer engineering categories. Under *Languages,* for example, there are groups that are interested in a number of programming languages: ADA, Visual Basic, C, C++, and others. The following sites are identified for C and C++.

---

comp.lang.c.moderated

- Moderator:
  Peter Seebach <clc>
  request@solutions.solon.com>

- Discussion List:
  info-c@arl.army.mil

- FAQ available:
  ftp://rtfm.mit.edu/pub/usenet-by-group/comp.lang.c/

comp.lang.c++

- Discussion List:
  help-c++@prep.ai.mit.edu

- FAQ available:
  ftp://rtfm.mit.edu/pub/usenet-by-group/comp.lang.c++/

There are other groups for various topics in chemistry, energy, physics, biomedical electronics, civil engineering, control engineering, heating ventilation and air conditioning, lighting, manufacturing engineering, and other special interests. In the Usenet category sci., the following newsgroups are listed for engineering:

sci.engr.	**sci.engr.manufacturing**
sci.engr.biomed	**sci.engr.marine.hydrodynamics**

sci.engr.chem	sci.engr.mech
sci.engr.civil	sci.engr.metallurgy
sci.engr.color	sci.engr.safety
sci.engr.control	sci.engr.semiconductors
sci.engr.geomechanics	sci.engr.surveying
sci.engr.heat-vent-ac	sci.engr.television.advanced
sci.engr.lighting	sci.engr.television.broadcast

DejaNews was created to track newsgroups, and it will search among several thousand Usenet newsgroups. The engine is very up to date and posts 500 MB of information a day.

DejaNews
http://www.dejanews.com/

Listserv Mailing Lists

When you open a Usenet group site, you investigate the material at your leisure. You can also place yourself on the mailing lists of groups, the most popular of which is Listserv. Messages will be forwarded to you on a regular basis if you subscribe, and you can "un-subscribe" at any time.

Civil engineering research and education	**civil-1**	**listserv@unbvm1.csd.unb.ca**
Computer-aided engineering design	**caeds-1**	**listserv@listserv.syr.edu**
Engineering and public policy discussion list	**eppd-1**	**listserv@unbvm1.csd.unb.ca**
Engineering societies	**engsoc-1**	**listserv@listserv.unb.ca**
Mechanical engineering	**mech-1**	**listserv@utarlvm1.uta.edu**
Minority engineering recruitment and retention	**merrp-1**	**listserv@uicvm.uic.edu**
National Society of Black Engineers	**nsbeline**	**listserv@listserv.syr.edu**
Science engineers list	**sci-eng**	**listserv@gu.uwa.edu.au**

Society for Advancement of Materials and Process Engineering	sampe-1	listserv@psuvm.psu.edu
Society of Mexican American Engineers and Scientists	maes-1	listserv@tamvm1.tamu.edu
Steam generators, piping, and equipment	steam-list	listproc@mcfeeley.cc.utexas.edu
Transportation science	tss-list	listserv@mitvma.mit.edu

Research Databases

There are databases under the administration of research organizations and professional societies. Access may be restricted, or there may be fees.

Advanced Technology Program
http://cos.gdb.org/best/stc/atp.html   (restricted access)

The preceding site is focused on U.S. business in technologies and is the work of the National Institute of Science and Technology.

Canadian Community of Science
http://cos.gdb.org/work/best-dbs-canada.html   (restricted access)

A database focused on research from Canadian universities.

CEDB-ASCE Database for Civil Engineering
http://www.ascepub.infor.com:8601/cedbsrch.html

This database is intended to provide information on all publications of the American Society of Civil Engineers (ASCE).

NASA Astrophysics Data System (ADS)
http://adsabs.harvard.edu/

NASA Databases
http://galaxy.einet.net/hytelnet/NAS000.html

NASA Technical Report Server
http://techreports.larc.nasa.gov/cgi-bin/NTRS

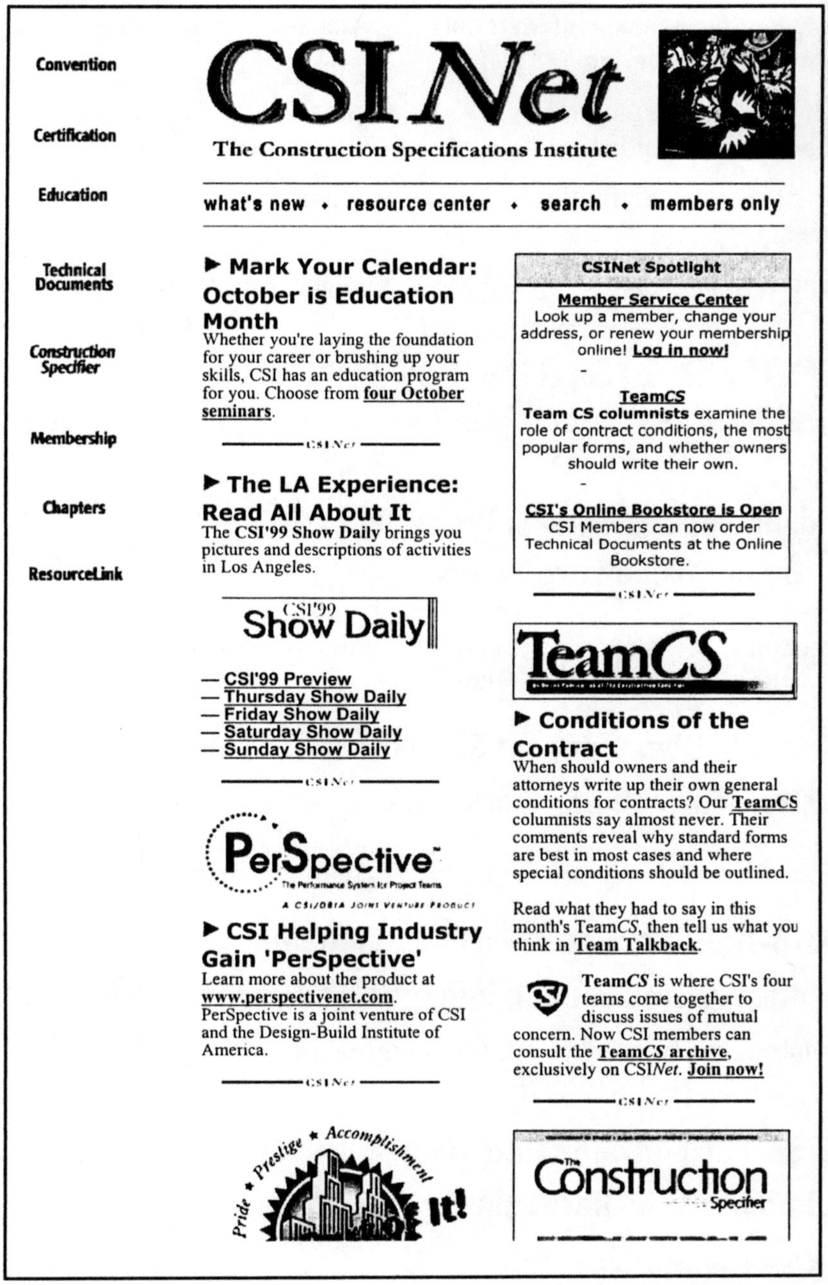

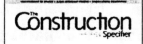

Figure 10-14 Construction Specifications Insitute Home Page. (Illustration Courtesy of CSI at www.csinet.org.)

The first NASA address is based at Harvard University and contains abstracts of nearly one million documents concerning astronomy, astrophysics, instrumentation, physics, and geophysics. The NASA Technical Report Server (NTRS) is an open access to the files of technical report databases maintained by NASA. Two million records cover all areas of engineering. There is no fee.

National Science Foundation
http://www.nsf.gov/

The NSF is an independent agency of the U.S. Government. The database includes a number of scientific areas including *Computer and Information Science and Engineering*. *Engineering* is a separate category.

Your Engineering Bookmarks

As you can tell from looking at the many Internet addresses that are included in this discussion, there is something on the Web for every engineer or engineering technician. Far from being daunting, access to interesting and valuable resources is easily undertaken, and there is no need to be confused by the complexity. For example, if you are searching for one or two names, in a phone directory, there is nothing confusing about the fact that there are 300,000 others in the book. Once you identify a few WWW addresses of interest to you, the particular focus of your professional field will yield results that are highly specific.

Since most of the Web is designed to be hierarchical, any category of interest quickly divides into subcategories. In this environment of reductions and divisions, subjects become smaller, more precise, and very manageable. As you find your way from one cyber location to another, you will close in on the Web sites you need to have available in your bookmark file.

Suppose you are a student in structural engineering with an interest in the construction industry. You will quickly realize that it may be of value to file the Web address of the American Consulting Engineers Council:

http://www.acec.org/

You also will want to have access to the American Society of Mechanical Engineers at

http://www.asme.org/

●

"The World Wide Web Site designed for the Professional Building Industry"

●

PREVIEW

● <u>**Real Estate Opportunities**</u> ᵁᴾᴰᴬᵀᴱᴰ

● <u>**"Frontier Dimensions In Building"**</u>

    • **"Sustainable Architecture -The Company"**
        ◦ With: *Lawrence Schecter - Architect, Land Planner, Author*

    • **"Soul and Earth"**
        ◦ With: *James T. Hubbell - Designer, Artist, Author*

● **Special Topics:**

    • <u>**"Visit Home Aid America"**</u>

● **Building Code Library: "ADA Access Guidelines"** (*Coming Soon*)

● **Utility Companies:** <u>**"Utilitly Companies Information"**</u> ᴺᴱᵂ

● **Environment:** <u>**"Green Building"**</u>

● **Building Disciplines & Services:** <u>**"Construction Resources Locator"**</u>

● **We Appreciate Your Input:** <u>**"Building Areas of Interest - Input Form"**</u>

● **Free Listings For New Projects:** <u>**"Building Projects Open For Bid"**</u>

● **Building Industry Publications:** <u>**"Automated Builder Magazine"**</u>

●

● **THE CONTENTS OF THE BUILDING SITE NETWORK have been compiled to provide the most extensive on-line assembly of building reference and research tools. Industry professionals and building information researchers, alike; find TBSN's "single resource" and "one-stop-shopping" design easy to use as well as "continually useful".●**

<u>Table Of Contents</u>

Figure 10-15 Home Page of the Building Site Network. (Illustration courtesy of BSN.)

With a little reading you will encounter other organizations that are important to you:

American Wood Council	http://www.awc.org/
National Association of Home Builders	http://www.nahb.com/
Associated General Contractors	http://www.agc.org

I stated earlier that standards and codes are very much a part of engineering design and application. In the case of construction engineering, for example, there are many Web addresses that are of value.

Construction Specifications Institute	http://www.csinet.org
National Institute of Building Sciences	http://www.nibs.org
Council of American Building Officials	http://www.cabo.org

One Web site should be an excellent source for every detail of the codes in the construction industry. The Building Site Network at

http://www.buildingsite.com

plans to contain the regulations for the following national standards:

> **National Electrical Code**
>
> **Uniform Building Code**
>
> **Uniform Fire Code**
>
> **Uniform Mechanical Code**
>
> **Uniform Plumbing Code**
>
> **Uniform Zoning Code**

It is in this specific way that Web resources can be of help. They are not difficult to locate once you have a resource such as this chapter. This is your starting point.

Internet Guides

Most of the Internet addresses in the preceding discussion were gathered from a very few sources that are of interest to engineers and engineering technicians. Although there are many books about the Internet, an industrial engineer or a network systems designer has very specific needs. You can, of course, use the Internet without a guide, but as you can tell from the addresses that appear in this discussion, it is very helpful to have a carefully selected set of resources as a starting point. My point of reference was men and women in engineering technical fields, so the presentation is very selective. If you would like to see more detailed discussions of interest to engineers, the following books are excellent resources and were important resources for information offered here:

Doherty, Paul. *Cyberplaces: The Internet Guide for Architects, Engineers and Contractors.* Kingston, MA: R.S. Means Co., 1997.

He, Jimin. *Internet Resources for Engineers.* Melbourne, Australia: Reed International Books, 1998.

O'Donnell, Kevin, and Larry Wright. *The Internet for Scientists.* Amsterdam, Netherlands: Harwood Academic Publishers, 1997.

Thomas, Brian. *The Internet for Scientists and Engineers.* Bellingham, WA: International Society for Optical Engineering (SPIE), 1997. (This text is co-published with two other societies: IEEE and the ASME).

Exercises Chapter 10, Section B

Complete one of the following exercises:

1. Use any one of the following electronic sources and explore the source in terms of your technical specialization. Copy a few index pages or home pages of interest and briefly investigate the resource as a potential tool in your work.

   - Libraries
   - Trade magazines and society journals
   - Professional associations
   - Engineering standards
   - Usenet groups
   - Listservs
   - Research databases

2. Using all the resources from the preceding list that you think will be of use to you, develop a brief discussion of sites you discovered among those resources. Briefly survey what you found at a few of the sites that were particularly interesting. Develop a memo to your instructor as in Exercise 1.

Develop a short discussion (up to 500 words) and format the project as a memo to your instructor. Include your screen scans as attachments.

Entering the Pipeline

The resume is possibly the most important single sheet of paper you will ever design. Given its role, you would expect colleges and universities to dedicate some small space in their programs to resume writing, but it is a rather awkward item to position in the normal scheme of things. Fortunately, placement services on most campuses often offer workshops to help students create an effective resume. The problem is that you must be made aware of the workshops, and the promotions for these campus activities often seem to fall through the cracks. Because the workplace communication volume of the *Wordworks* series is concerned with such activities as job interviewing, it is important to dedicate a discussion to resume writing. Since the *Writer's Handbook* has a strong focus on writing procedures of interest to engineering technicians and engineers, it seemed most appropriate to conclude with a chapter concerning resumes, with specific reference to engineering technologies.

11

Sample 11.A

Michael Late
1407 W 184th
Seattle WA 98002
206-365-0000

OBJECTIVE: To seek work as an entry level technician.

EDUCATION: Associate of Applied Science Degree in

Electronic Engineering Technology from North

Seattle Community College.

FURTHER EDUCATION PLANS: Bachelor of Science in

Electronic Engineering Technology from Western

Washington University.

Clubs: Member of Phi Theta Kappa (National Honor

Fraternity for Community Colleges)

    National Dean's List

    Member of IEEE

Hobbies: Hunting and Fishing

Given the small size of the document, the resume assumes a surprisingly large role in the job placement activities of college men and women. It is, of course, nothing more than a *history,* but for that very reason it has become the most practical tool you can develop for job hunting. For potential employers, it is also the most practical tool they can use for an initial quick measure of your appropriateness for a position. For both the employer and the potential employee, the resume is a compact and valuable introduction.

You might think that you have little to say in a resume and very little space in which to say it. True. Without some guidance, a resume could be little more than a sketch—a silhouette at that. It is important to realize that the resume is a highly defined profile that has a mission. It is a history, but it is a *marketing tool* as well. If it is handled as a sketch, the product may look like the brief resume by Michael seen in Sample 11.A. If it is designed with care and motivation, perhaps because you really want that certain job, the resume will look more like Rand's project (Sample 11.B). The challenge is to take whatever information you can gather and construct a top-notch resume that gets attention.

Horror stories about 400 applications for a job are not that outrageous. A student was recently skunked on his admission hopes for a university engineering program because the program admissions committee accepted only 35 out of 900 applicants! In cases such as this, the odds are against the applicant, but even if there are only ten other applicants, you cannot overlook the fact that the odds are still 10 to 1. You must hope that every point you can muster is one more point than the competition is scoring. It is important to see the resume as a contributing factor in determining the points that are scored.

Yes, the employment or school history will tell the tale, but there is another way to see a resume. If, as often happens, a great many applicants or candidates meet the basic requirements, then the measuring tools for selection redefine the rules of the game. This situation often occurs at major institutions—when excessive numbers of applicants meet the entrance requirements at major universities or when excessive numbers of people pass a civil service entrance examination for city, state, or federal agencies. In these common situations, applicants are then evaluated by a great many other measures, and, at this level, the resume can take on a renewed mission.

You must understand the *mission* of the document first and try to get beyond the overly simple idea that the resume is little more than history.

Sample 11.B

(206) 244-0000

3394 Princeton Blvd.
Tukwila WA 98188

RAND SCHINDLER

OBJECTIVE

To obtain a position leading to responsibilities involving engineering or marketing in the biomedical electronics field.

EDUCATION

1997–1999: Attended North Seattle Community College, acquiring 140 credits of solid technical schooling, resulting in an A.A.S. degree in biomedical electronics.

MAJOR TECHNICAL CLASSES

Biomedical Electronics I, II, III	Integrated Circuits & Adv. Solid State
DC Principles of Electronics	Programming in Basic
AC Principles of Electronics	Electronics Drafting
Solid State Fundamentals	Digital Logic & Comp. Fundamentals
Soldering & Handtool Practices	Technical Report Writing
Basic Circuits & Systems	Technical Physics
Electronic Troubleshooting	Anatomy & Physiology
Advanced Circuits & Systems	Medical Terminology

Also completed two years of liberal arts education at the University of Washington.

Maintained a 4.0 grade average at NSCC, was the recipient of the NSCC Merit Scholarship, and am a member of Phi Theta Kappa.

EMPLOYMENT HISTORY

1986–1997

Employer: U.S. Navy

Duties: Leading Petty Officer for the electronics maintenance division. Was in charge of 15 people. Also was the Assistant Special Services Officer in charge of 9 people during the same time period. Given a Top Secret Clearance.

Military Schooling

Schools: Ocean Systems Technician
                Electronic Maintenance
            Career Counseling
                Management and Counseling
            Basic Electricity and Electronics
            Sonar Systems
                Operator and Maintenance

REFERENCES

On request.

360

Exercises Chapter 11, Section A

Career Goals:

_____

Assess your career goals by developing a list of ambitions you would like to fulfill. It will be helpful to make an initial list that concerns employment goals alone. If the list is difficult to construct, think in other terms such as your interests, your values, your self-expectations, or your social commitments. Under each goal or value or interest, explain the perception in a sentence or two.

_____

Special Skills:

_____

To the extent to which you can identify your unique skills, identify them. Be sure to add nouns to the action verbs.

Ø   Professional drafting experience . . .
     Professional drafting experience/AutoCAD Release 15

_____

Briefly describe each skill in some way.

Three years of experience with ABC Incorporated in commercial design applications for structural plumbing and electrical drawings. Completed drawings for twelve clients.

_____

These skills are not exclusively of a technological nature. Include supervisory and other employment-related considerations.

_____

_____

Design Sense and the Resume

Impact

The resume is a marketing tool. It is veiled as the historical account of your employment, but you are trying to *market* your skills or your learning or your experience. Whatever you plan to do with your history must be governed by a keen sense that this document is *not* simply itemizing a product. It is not simply an index to your employment or schooling. It is product *packaging*. The schooling or employment is inside the box. The resume is the box; it is a matter of merchandising. The first sentence of this chapter referred to *designing* a resume, not to writing a resume.

First, realize that employment applications are not created in your best interest. At their worst, they are generic applications purchased from a large business products firm. At best, they are tailored to meet the needs or standards of large corporations. In other words, the one, two, three, or four pages of questions certainly were not constructed with any one job or one person in mind. The resume, on the other hand, is constructed to meet *your* ends. There is a huge difference. You may think that the information you provide on an application and the information you provide on a resume are functionally identical; however, the *order* and the *priorities* of the two sets of historical information differ considerably because the documents differ in structure.

In the same sense that a corporation can design an application to suit company needs, you can design a resume to suit your needs. Your first consideration is layout. You must design the document to sell, and it must sell fast. The grim stories about the 400 applicants get worse when you see the selection committees scanning the supplemental documents. The resume is usually the most important of those, and it will get some attention—for a minute or two. In that length of time you must build interest. Of course, this is the worst-case scenario. Let's hope the parking lot is empty the day you apply. That could well be the case, also. For marketing purposes, however, you must be competitive and assume you need a quick and competitive edge. What should you do?

The first tactic is to prioritize the design. There is a standard formula used in newspapers that is well understood in all the advertising industries. The point of contact for the eye is *always* at the upper left on a page. We read diagonally down a page to the lower right.

Sample 11.C

Sharon Hiromi
DRAFTER
45482 W 56th St.
SEATTLE WASHINGTON 98020
PHONE: (206) 775-0000

CAREER OBJECTIVE:

Electromechanical, Industrial CAD Drafter or Designer for an Engineering or Manufacturing firm.

EDUCATION EXPERIENCE:

North Seattle Community College, 9600 College Way N., Seattle, Washington
      98103 March 1997–June 2000
    Degree Awarded: Associate of Applied Science in Industrial Drafting.
    Course Content: Mechanical Detailing, Isometric, Dimensioning and Tolerancing, True Positioning and Geometric Tolerancing, Descriptive Geometry, Sheet Metal Drafting, Schematic and Logic Diagrams, PCB Drafting, Piping, Welding, Gears, Documentation, ANSI and Mil-Standards.
    Additional Course Work: Technical Math (Basic Trigonometry, Pre-Calculus, Vectors, Algebra, Graphs, Ratios, and Proportions); Technical Physics (Types of Forces and Linear Motion, Work and Mechanical Energy); Technical Report Writing and Communication Skills.

WORK EXPERIENCE

DATE	EMPLOYER	POSITION
July 1990 December 1997	Lockheed Shipbuilding and Construction Co.	Engineering Aide Drafter

Harbor Island, Seattle, Washington

Duties Included: Mechanical and structural drafting using CAD, ink on mylar, and graphite on vellum. IBM and NEC AutoCAD systems work. Also as-built drawings and presentation graphics.

PORTFOLIO AND REFERENCES AVAILABLE UPON REQUEST

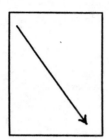

This is your design. Whatever you want to market must be on the top half of page one and *must* collect and focus a reader's attention. The often-quoted maxim that a resume "should be only one page long, or no one will read it" is hardly accurate. I argue that your resume has to be only half a page long or no one will read it! Of course, if the first half of the first page sells—*if* you get the attention of your reader—it hardly matters how long the resume is. What matters is the point of contact, where you position it, and what it holds. If you are familiar with the discussions of cover-page design work in *Technical Document Basics,* you will see a similarity. A resume is designed with the same above-center impact as a title page, as you will notice in Sharon's resume (Sample 11.C).

What goes into the target area is the next issue. For college graduates, the choice is usually obvious: education. For community college graduates, the choice is not as clear, since many junior college men and women have years of related work experience. Often they are simply returning to complete a degree because their careers have hit the wall standing in front of the next promotion: no degree, no raise. In cases such as this, the employment highlights might be more important to market. Occasionally, there will be yet other highlights that might replace education or employment. If someone has had success with some aspect of the computer industry, for example, those particular merits could be the focus of the resume, in which case the education and employment pictures could move down the page.

The Outline Style

The second critical tactic is also a layout matter. The topography of the document must be very clear. In other words, you must imagine a grid upon which you construct the

Sample 11.D

Harold Breen
1820 Colchester Ave.
Burlington, Washington 98334

(206) 553-0000

Seek employment in drafting and design in the architectural field. Have experience in mechanical, HVAC, and electrical drafting and design. Experience in all drafting fundamentals and basic computer-aided drafting. Have skills with standard and plastic lead on mylar or vellum and AutoCAD.

EXPERIENCE:
- 1996 to present—John Fluke Manufacturing in the Facilities Support Group. 6920 Seaway Blvd., Everett Washington 98206
(206) 356-2136 (Jim Novlan, Supervisor)

    Job Title/Duties: Detail Draftsperson/Engineer's Aide, full-time. Design architectural, electrical, mechanical, and HVAC additions and revisions within the facility. Coordinate projects with contractors and in-house maintenance. Responsible for organization and upkeep of approximately 1000 working drawings. Provide presentation drawings and visual aids for corporate meetings.

EDUCATION:
- 1995–1997 North Seattle Community College. Associate of Applied Science Degree in Construction Engineering Drafting, December 1997.

Program Core

Basic Drafting Fundamentals	Advanced Drafting
Construction Materials	Structural Systems
Applied Mechanics	Electrical & Mech. Systems
Construction Estimating	Reinforced Concrete
Detailing	English Composition
Adv. Contract Document Preparation	Technical Writing

- Renton Senior High
Renton, Washington—Graduated 1995
Major program core: Beginning Drafting, Intermediate Drafting, Advanced Drafting.

REFERENCES: Available upon request.

information. The resume is prioritized from top down, but it is also prioritized from left to right. Fundamentally, the resume is an outline. By tradition, the resume is handled without alphanumeric structures, but it is otherwise identical to the outline in the use of indented subsets. Subsections move in and down in a way that you should find familiar, since the *Wordworks* series concentrates on the importance of outlining.

Simply outline your historical material and then construct the resume without numbers or letters. Use boldfaced key terms to initiate the major sections, and order subsections with bullets (simple black dots, for example). Keep the grid concept in mind so that the vertical and horizontal planes define the content. Use the design of the document to help build a neat and orderly appearance. Allow for vertical and horizontal planes of white space to enhance the visibility of the logical sections. You should be able to hold the resume at arm's length and see the logic of its structure! You can do exactly that with most of the resume samples including Harold's (Sample 11.D). Do not cram the design. The outline should be suspended in the space on the paper. The use of white space is critical to the clarity.

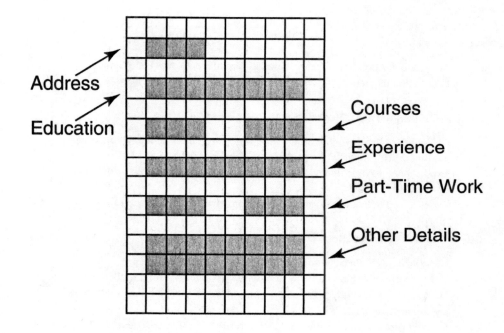

You can use headings in the document. Most resumes will utilize at least a few of the main headings we have already mentioned.

EDUCATION

EMPLOYMENT

Sample 11.E

TERRY WESTING
1883 BOTHELL ST.
REDMOND, WASHINGTON 98056
HOME: (206) 860-3456
MESSAGE: (206) 860-6543

PROFESSIONAL PROFILE

CAREER OBJECTIVE

To obtain entry in a growing electronics firm with an opportunity for advancement based on individual initiative and productivity.

EDUCATION

North Seattle Community College
9600 College Way N
Seattle Washington 98103

- Associate of Applied Sciences degree in Digital Electronics Technology.
- Graduation expected December 2000.

HARDWARE DESIGN AND TROUBLESHOOTING SKILLS

- Designed and constructed digital lock, traffic light controlling system and elevator control circuitry.
- Experienced in mechanical and electronic drafting according to ANSI standards.
- Experienced in troubleshooting televisions, tape recorders, and radios.

SOFTWARE DESIGN SKILLS

- ROCKWELL AIM-65,assembly.
- Wrote automatic chip testing system with interface to testing pad.
- Designed traffic light controlling system with interface to custom designed software.
- General programming experience with INTEL 6502 microprocessor, HP3000 PASCAL, APPLE II BASIC.

OUTSIDE INTERESTS

Astronomy, Navigation, Boating, Cycling, and Cooking

REFERENCES AVAILABLE UPON REQUEST

These in turn may be accompanied by appropriate subheadings:

Dates   Address   Supervisor

Terry's resume (Sample 11.E) is quite modest in content. She used blocks of outlined information, simple and spacious topography, and striking headings to make the material impressive.

As you begin to build the resume, be thinking about the final appearance as a design consideration. The overall design poses a particular problem: the resume must gain attention but without much fanfare. It is difficult to make a resume stand out. You want the top half of the first page to communicate quickly; otherwise, it is hard to maneuver your resume to the top of the pile because it is a very traditional tool that is modeled in a conservative fashion. Resumes can be overdone, though. I was once told that a resume I designed for someone was "too slick." I still have that resume; it was simply ahead of its time in terms of the computer applications that were used to build it. The supervisor who thought it was overdone still had an electric typewriter in her office. Nonetheless, there is a message here that you do not want to overlook. A resume can be too catchy or a little too close to the cutting edge. I have several beta samples from Microsoft in multicolored fonts. They were never published for obvious reasons.

As tempting as those novelty paper stocks might be at the copy shop, you cannot print your resume on a sheet of blue sky with a puff of clouds. You can use pastels or any high-quality paper. Heavy papers are ideal, but do not use card stock. You can also use your own letterhead paper if your resume is being used as an entrepreneurial tool for your own work or for proposals or portfolios. I designed one for my work that had a removable business card in the letterhead position at the top. The idea was that it could be removed and positioned in a client's card file. Resumes are clumsy in this respect, and a removable card was a convenience—as well as my best kept secret for adding novelty to a resume. It was a unique idea that always got attention.

A few other devices that are attractive include a border or perimeter or some other design touch, such as a logo. A little use of color is also a reasonable option, but remember to keep the resume very conventional overall. CAD-generated resumes will draw attention, as you will notice in Sample 11.F, but you would use CAD only if the situation merited the style. Whatever little touch you invent can add interest. Be inventive but cautious. We experimented with tables on Bob's resume on p. 404, and the result was unique as well as attractive. Your simplest option is obviously your choice of a font that will have an attractive—but appropriate—appearance. There are at least three thousand on the market, and you can buy one on a disk and create a more or less personalized product.

You can also consider using software templates. *Microsoft Word 6.0* (and subsequent versions), for example, has a wide variety of resume configurations that are divided into four different *styles,* from entry level to professional. Certainly, this sort of a tool is an excellent way to produce an initial draft of a resume, and there is no reason not to use templates.

Sample 11.F

JEFF BUCHANAN	4012 6th Ave W
	Seattle WA
EMPLOYMENT OBJECTIVE	98108
POSITION AS A CADKEY/CATIA OPERATOR	(206) 555-0000

QUALIFICATIONS

Working in the field of Computer-Aided Mechanical Drafting
as a CADKEY and CATIA operator for the last four years.
Familiar with the drafting standard of the Boeing Commercial
Airplane Group.

WORK EXPERIENCE

'B' Grade drafter at Boeing Commercial Airplane Group.

Teaching assistant, North Seattle Community College,
CAD LAB. Mechanical and Structural.

ELECTRO-MECHANICAL DRAFTING SKILLS

Familiar with the Geometric Dimensioning and Tolerancing
Standard per ASME Y14.5 1994
Familiar with honeycomb core composite assemblies.
Plastics fabrication (fiberglass & thermoplastic).
Metal Fabrication (sheet stock & aluminum extrusions).
Generate schematics net-listing for PCB design.

COMPUTER SOFTWARE EXPERIENCE

- CATIA
- CADKEY 1997
- APLG (Automated Parts
 List Generator)
 AutoCAD - Release 14

- DIOL (Drawing Index On-line)
 Word Perfect
 MS-DOS
- Microsoft Word

EDUCATION

NORTH SEATTLE COMMUNITY COLLEGE
9600 College Way N., Seattle WA 98103
Associate of Applied Science Degree
Electro-Mechanical Drafting 2000.
BALLARD HIGH SCHOOL Seattle, Washington
Graduated 1996

ADDITIONAL COURSES
- Math through Calculus
- Advanced MS-DOS
- Technical Writing

REFERENCES AVAILABLE UPON REQUEST	SCALE 1:1	SHEET 1 OF 1

However, there might be similar-looking resumes in the pile. The point of this chapter is to suggest ways to be or look original. You are going to have enough problems competing with all the other graduates who have the same training. The similarity of professional backgrounds can make the choice among a group of finalists quite complicated. You certainly do not want your resume to look similar as well.

> **Use italics for emphasis.**
>
> **Use boldface for headings.**
>
> **Use underlines to highlight.**
>
> **Use drawn lines as needed.**
>
> **Use bullets to help clarify lists.**
>
> **Use parentheses where appropriate.**
>
> **Use a distinct font for the name and address at the top of resume (if desired).**

E-Resumes

It is interesting to note the trend toward prefigured resumes that are designed to be scanned by computer. As an applicant, you might be required to fill out the *company resume* so to speak. For its part, the company can then sort the resumes by desired employee criteria and produce a list of acceptable talent. In effect, an electronic resume might be better called an electronic summary.

A resume is a carefully crafted marketing tool. The electronic summary eliminates all the craftwork so that the document will be "system friendly." The document has to be simple and tidy so that it can be read by optical character recognition (OCR) software. The entire document is left-justified, and you are supposed to follow guidelines such as these:

> **Do not use more than one font.**
>
> **Do not use italics.**
>
> **Do not use boldface.**
>
> **Do not use underlines.**
>
> **Do not use drawn lines.**
>
> **Do not use bullets.**
>
> **Do not use parentheses.**

As you can see, the guidelines for electronic resumes are constraining and largely defeat the popular design concepts for the resume. In fact the best possible guidelines for an effective resume are quite the opposite.

What to do?

If you send an electronic summary and succeed in gaining an opportunity for an interview, take your hardcopy resume with you and give it to the parties involved, because the electronic summary is certainly not a carefully crafted resume, and the two tools should not be confused.

Exercises Chapter 11, Section B

Education:

Outline your educational history: identify all institutions that have contributed to your post-secondary education. These include technical schools and private colleges, community and junior colleges, military schools, and universities. Develop a list of the schools in reverse chronology. Use the three-column format and identify years, institutional addresses, and the program of study.

Create a list of subjects studied at each school and prioritize the course work. Briefly explain the importance of each in terms of your learning path. Focus on the relevance of your present learning experience, but explain relevant education of any sort.

This document can be kept in you portfolio in case an employer requests a specific profile of your course work and the content of your academic programs. The lists of courses and the brief descriptions are more easily understood than the information provided on transcripts from institutions. In addition, the course-work profile serves as a practical reminder you can review prior to an employment interview. Selections can be taken from the profile and used in cover letters for the resume or application letters for employment (which are often identical).

Sample 11.G

Michael O'Hara
6738 Sierra Lane N
Seattle WA 98103
(206) 301-000

PERSONAL OBJECTIVES:

- Electronics Technician, Energy & Process Control Technician
- C.N.C. Technician
- CAD Designer

PREVIOUS POSITION:

Electronics Technician II, Renton Vocational Institute, 1990–1995.

Duties:

Calibration and Repair of Electromechanical-Servo systems, Programmable Logic
   Controllers, AC and DC Variable Speed Drives, Hydraulics and Pneumatics.
Organization and Distribution of Reference Materials, Parts, and Supplies.
Aid Students and Instructors in Project Design, Development, and Demonstration.
Begin with Layout, Fabricate, and Test Circuits.

SUMMARY OF QUALIFICATIONS:

Six months electronic lab assistant at Renton Vocational Institute.
Six-month internship in electronics and office machine repair with the Renton School District.
1000+hours of hands-on Electronics lab experience.
1000+hours of hands-on Energy & Process Control lab Experience.
500 hours CAD lab experience.

EDUCATION:

AAS Degree in Energy & Process Control at Seattle Community College, June 1999.
AAS Degree in Electronics Technology at SCC, August 1998.
Two quarters of AutoCAD, May 1996.
Maintained a GPA of 3.70 and was president of Phi Theta Kappa honor society for SCC.

ENERGY AND PROCESS CONTROL EQUIPMENT OPERATED, CALIBRATED, AND REPAIRED:

AC & DC Drives	Motor Control Systems
AC & DC Three-Phase Motors	Hydraulics & Pneumatics
Magnetic Starters	Tranducers & Sensors:
Programmable Logic Controllers	Panel Instrumentation (Optical, Pressure)
Single-Phase & Three-Phase Transformers	

ELECTRONIC EQUIPMENT OPERATED:

Computers & Printers	Current Tracer	Analog & Digital Multimeters
Software	Synthesizer	Microprocessors: Intel Family
Oscilloscopes	IC Tester	Capacitor & Diode Analyzer

ELECTRONICS CALIBRATED & SERVICED:

Computers	Printers	Industrial Microprocessors
AC & DC Drives	Printing Calculators	Oscilloscopes
Network systems	Disk drives	

REFERENCES: Available upon request.

Resume Contents

As observed earlier, the resume is probably the single most important project you will ever write. It is perhaps for this reason that this simple, conservative page or two of writing is peculiarly controversial. Many well-meaning well-wishers will gladly tell you how to go about it. The problem, you will quickly realize, is that no two of them agree. You will soon tire of the conflicting descriptions of a "good resume."

A few of the more prominent perceptions about resumes are as follows:

- A resume must be no more than one page, or no one will read it.

- It had better be an attention-getter.

- Work experience should be listed from past to present.

- Your college courses must not be listed.

- Your G.P.A. must be included.

- You cannot include a personal section anymore. It is illegal.

- You do not ever list references.

These and other ideas are common. Since everyone has his or her own thoughts about how the job should be done, let's strike a bargain. I, too, think I know how it should be done. What you need to do is agree to hear me, and then hear out all those other ideas that will contradict what you read here. But, as part of the bargain, you must agree to ultimately take the best ideas from all of us and do what *you* want to do. You see, there are no rules. A lot of people will volunteer their ideas, but, in the end, the selection of the content and the organization you build for the document are considerations that are in your hands. None of the suggestions you are offered are right or wrong as such. The product is much more flexible than any one of us probably realizes, as you can see as you examine Mike's energetic resume (Sample 11.G).

A number of typical features are commonly found on resumes:

- An objective
- Special skills or qualifications
- Education
- Employment
- Military history
- Computer know-how
- Awards and certifications
- Personal interests
- References

Other sections can be developed as well, and you certainly can develop far fewer if you choose to do so.

Sample 11.H

James Becker
Construction Supervisor

4020 20th NE
Seattle WA 98103 206-336-0000

CAREER OBJECTIVE: Construction Administration where experience in architecture, engineering, and review processes will help to ensure project success.

QUALIFICATIONS: Progressively responsible positions demonstrating success in

On-site Management Skills	Code Analysis & Construction Problem Solving
Technical & Administrative Decision Making	Team Relations

RELATED ACHIEVEMENTS: Analyzed data and developed information that assisted the Association of Building Officials in the development of new Code Standards.

Prepared technical reports and presented land-use data to the Portland Planning Agency that helped expedite the decision-making activities.

Proposed, obtained, and administered grants for Portland municipal projects that resulted in improvements in city services and the reconstruction of several municipal projects.

PROFESSIONAL EXPERIENCE: Assistant Engineer, City of Portland
Oregon Building Inspector, Salem, Oregon
Production Draftsperson, D. J. T. Designs, Beaverton, Oregon

EDUCATION: B.S., Construction Engineering, Oregon State University
A.A.S., Construction Technology, North Seattle Community College

CERTIFICATIONS: Council of American Building Officials Certification
International Conference of Building Officials
Building Inspectors Certifications

REFERENCES: Available upon request

I will discuss each category in turn, from the perspective of community college men and women in engineering transfer and engineering technology programs. In fact, the discussion specifically concerns graduating students. If you decide to prototype a resume, remember that it is easier to design it as though you have already graduated. You cannot, of course, use such a resume until you graduate. If you need an interim resume, for summer work for example, you can build one that fits your current status.

Position Desired
or Qualifications for a Position

There are basically four ways to begin the resume. You need to indicate immediately what you *want* or what you *do*. In a very few seconds, the document must indicate the employment picture. This can be handled in several ways:

- State an objective

- List qualifications

- List skills

- Identify your official title

James's resume (Sample 11.H) is unique in using three of the four in the same document. This resume opens in a rather complicated fashion, but you probably need nothing as forceful as this. The graduating student should probably state an objective and be done with it. You will notice that most of the resume samples in this chapter state the employment goal in simple terms.

Objective: Seeking a position as . . . .

There are other alternatives you will want to consider later.

Once you have gained experience in your field of interest, you can remove the objective and add a list of qualifications or a list of skills. Prior to your gaining significant work experience, I would not encourage you to use a list of skills—first, because this might not be possible or, second, because it might tend to promise more experience than you can deliver. Once you have an employment history that *demonstrates* your qualifications or skills, the resume that identifies such skills will be more convincing. These highlights can be listed at that time. Like the objective, a brief list of skills or qualifications will open the resume with the immediate employment picture.

Another device—the professional title—is equally effective. If you are a licensed architect or engineer or engineering technician, you can simply identify yourself by your title. You may add a license number if you wish, although this is more commonly placed on a busi-

Sample 11.I

PETER RICE
Construction Cost Estimator

356 Meridian Ave. N
Seattle Washington 98103
(206) 781-0000

EMPLOYMENT

Sigma Construction:	**881-0000**
1997–Present	Full time

Estimator:
  Responsible for material take-off, pricing, compliance to specifications, and completion of proposed bids.

Tullson Brothers:	**883-0000**
1995–1997	Temporary, Full time

Estimator:
  Responsible for material take-off, pricing, compliance to specification, and completion of proposed bids.

Engineer's Aide:
  Home office assistant to field superintendents, ordering needed material and arranging for delivery to various job sites.

S.J. Corrigan Inc., Inter-Media Division:	**365-0000**
1990–1995	Part time

Interior Drafting Designer:
  Preliminary interior design of professional offices.

Draftsman:
  Drafting finished drawings for company clients.

Northwest Computer Service:	**237-0000**
1985–1990	Full time

Computer Graphics Draftsman:
  Responsible for drawing quality and accuracy. Proprietary

EDUCATION

Degree:	University of Montana
	B.A., Construction Engineering
Honors:	University of Idaho,
	Design Award:
	Design & Construction of Office Facility, Boise, Idaho

REFERENCES

Mr. Scott Koopman	Project Supervisor
	Wk: 483-1193 Hm: 822-0000
Ms. Karen Anderson	Structures Professor, U of O
	Wk: 543-9223 Hm: 682-0000

ness card, if it appears at all. Notice the title in Peter's resume (Sample 11.I). Peter simply refers to himself by his trade: *estimator*. By using your formal title, you essentially replace the other devices with certain professional expectations implied by your position. You show that you have a seal of approval. Of course, there are no rules, as I said. You can have a title, an objective, *and* a list of qualifications, although this may slow down the reader. One of the four conventions is sufficient, and the objective is the most practical for the new graduate.

Although the objective or the list of qualifications is at the top of the page, your name, address, and phone number are uppermost. Use left, center, or right locations for the address. Be sure there is a functioning daytime phone number because personnel officers will not be calling after normal business hours. Arrange for voice messaging. Do *not* list your work phone. Current and potential employers consider this bad form.

Education

If you do not have a strong background of appropriate employment, then focus on marketing your education. Both areas are of equal concern to employers, so develop your strongest position first. Both education and employment are *usually* listed in reverse order—from the present to the past. Generally, your current status is your best selling point, so the reverse chronology usually works. Begin with your most recent college experience and enter the information that follows.

School (inclusive dates)

Complete address

Phone number (registrar's office)

Degree (do not abbreviate)

Specialty

Graduation date (you can also say "expected date")

Program of studies (highlights)

You can see from the various resume samples that the layout of this information can be handled in a number of ways. Let me anticipate your likely questions. Add other colleges that you attended for a quarter or more if the grades were reasonable. If you bailed out of one, you can probably be forgiven for the oversight of omitting it. Resumes must be truthful, of course, but they are not always thorough or inclusive. Also, if you went to an embarrassing number of schools in false starts, you can omit a few. The idea is to collect the valuable experiences for a prospective employer to see. In particular, do not omit college work because it is *unrelated*. All college work represents a certain professional capability. A year here and a year there all add up.

Sample 11.J

Edward Quintero

Job Objectives

Seeking a position in the electromechanical field: service, maintenance, or technical support.

Education

1998–2000   Associate Degree in Applied Sciences; concentration in Electrical Power and Control Technology. North Seattle Community College, 9600 College Way North, Seattle, Washington 98103.

Major Courses:

Electrical Theory, DC & AC	Static Control Systems
Solid State Fundamentals	Electrical Code Design
DC & AC Machines	Ladder Logic
Digital Logic	Basic Circuits/Systems
Instrumentation & Servo Systems	Hydraulics, Pneumatic Systems
Program Controllers	& Controls

Support Classes:

Technical Math for Electronics
Technical Physics
Technical Report Writing
Electronics Drafting
Soldering/Handtool Practices

1978–1998   Various Military Training Schools: Personnel & Management Schools, Radar School, Racial Relations & Mediation, and numerous theory classes.

Work Experience

1978–1998   Active service in the United States Navy as supervisor, administrator, liaison officer, cost estimator, and teacher. As liaison officer, I coordinated functions between the Navy and the public. I am used to working with people and the high pressure of last-minute details or difficulties. As an administrator, I determined cost estimates for numerous presentations, implemented in-house educational classes, and wrote personnel evaluations.

Personal Data

Service:	Veteran of foreign wars, 20 years active duty, top secret clearance
Bilingual:	Spanish and English
Health:	Excellent; last physical exam, 1999
Hobbies:	Cooking, refinishing furniture, piano, growing roses
Residence:	Would relocate for proper opportunity
	8549 NW 15th
	Seattle WA 98117
	(206) 860-3838

References available on request

For uncompleted programs you can simply identify the work as

> Two quarters of general studies,

or you might say

> First year, pre-engineering program.

Include the years. Include the month of graduation if you completed a degree. GPAs and Dean's List awards are usually omitted. This is always surprising to new grads who have just fought long and hard for that 3.6. You *may* put the GPA or academic honors in the resume. Later you will take them back out once your work history assumes more importance.

Do list your course work as you see in Edward's resume (Sample 11.J). Prioritize the list. Generally the last courses will, therefore, go first. This list is intended to market your technical knowhow, so leave out the P.E. courses, the social science electives, and similar material. List what the employer might find important. If you are told not to list the course work but to give the prospective employer a transcript instead, consider the realities. Transcripts are coded, they are in fine print, they are hard to figure out—and they are *not* prioritized by your needs. They are also in historical order, which is not the format of a resume. And, finally, the *official transcript* is not official in any case if you take it to an interview. List the courses in order of importance, *without* institutional numbers, *without* abbreviations. If you say

> EET 201, 202, 203,

it will not be as meaningful as saying

> Electrical Engineering: Digital Circuit Design (three quarters).

You might also be wondering about other education, such as in-service training, company seminars, military programs, or other educational experiences. (College men and women usually omit high school education unless it was strongly technical or otherwise unique.) Feel free to list everything you want as long as it is relevant and as long as you are not losing the reader's interest. Remember, there is more to the resume than the education section. If you have extensive certificates (which you should *always* save by the way), they can be itemized on a separate sheet of paper that acts as a supplement to a resume. These certificates are quite common and, in a supplement, are very helpful in specialized fields such as Fire Command Administration.

Sample 11.K

Robert Kangas   102 Main St., Apt F
Northland WA 98011-3402
(206) 489-0000

OBJECTIVE

To obtain full-time, permanent work where I can best utilize my skills in integration and installation of PCs, troubleshooting, and repair.

EXPERIENCE

1997 to present	Design, building, and installation of control systems using Pentium-based computer systems. Technical Support at Phototronics Corporation, 5110 190th Street N/W, Suite 100, Lynnwood, WA 98035 (206) 776-0000.
1996	Integration and installation of IBM AT, 486, and clones for use with DOS.
1993 to 1995	UNIX office systems. Including cabling of terminals and *DXD* Computer systems, Lynnwood, Washington.
1990 to 1993	Repair, maintenance, and installation of word processors, printers, and scanners. Field Service Technician at Proline, INC., Seattle, Washington.
1986 to 1990	Customer Service Representative at Singer Business Machines/ Seattle, Washington WRT Customer Service Division, Bellevue, Washington.

EDUCATION

North Seattle Community College 9600 College Way North Seattle WA 98103 (206) 527-0000	(April 1999) Associate of Applied Science Degree in Electronics Technology. Solid State, Digital, and Analog Electronics Troubleshooting, Industrial Applications, Discrete Circuit Analysis & Applications, General Physics, Calculus, Technical Writing.
Singer Business Machines / WRT	Company schools: Point-of-Sale Terminals and Docutle Automatic Teller Machines.
U.S. Air Force 1981 to 1985	Electronics, Digital Electronics, and Computer Maintenance.
Meadow High School Renton Washington.	Electronics and Computer Math.

INTERESTS

Amateur Radio   KB8WNK

References provided upon request.

Since you will probably use a heading for the education section, you can simply develop an additional heading for *Other Education*. This is a common procedure. In fact, you can subdivide education to suit your needs. Suppose a student has a recent liberal arts degree, an earlier biomedical electronics degree, and now wants to go into tech writing. Because reverse orders are taboo, the solution is to create two sections—*Related Education* and *Other Education*—and then move the biomedical degree up the page to the top. You may break the chronologies to shift the reader's focus of attention by designing categories that allow you to move the material around.

Employment

The employment history is the other major part of the resume. If your employment is largely unrelated to your future plans, you can use the simple three-column format that you see on Sharon's resume (Sample 11.C). This is a convenient way to list employment without detailing the material. Do not omit employment simply because it is unrelated. You need to show that you have a demonstrated record of employment regardless of the type of work performed. However, you do not have to list *every* job you have held, particularly short stints or positions you would rather forget. For example, many women omit clerical activities from their employment history for fear that they will be invited back into those capacities.

Omit months of employment, and round out the jobs in years only. This practice looks more attractive on the resume, and it avoids specifics about lesser jobs. Do not identify positions by civil service numbers or company job titles that a reader will not understand. You should say "installer" and not "J-4." Also, if a position was temporary, indicate so by saying "part-time," "seasonal," "summers only," or some such explanation.

A background of professionally interesting employment calls for more details. Career-related employment makes you a serious contender for a position, so you must market the product. Supplement the preceding information with a sentence or two that describes your last position. Notice Robert's strategy in Sample 11.K. Do not bother to identify your earlier ranks and tasks within the same company unless they are relevant. You may write as much as a paragraph for a job description. Because you do not want the resume to be too long, do not detail more than two or three positions in this way.

Sample 11.L

Teresa Stimson

Project Engineer for Mechanical Systems and Energy Conservation Incorporated

As a Project Engineer for Mechanical Systems, my responsibilities included design and specification of HVAC, plumbing, and fire protection systems from schematic design through construction administration. Experience includes schools, housing facilities, and office buildings with energy conservation design specifications on many of the projects.

Representative Projects

Energy Upgrades (Energy Conservation Design Studies)

Kamiak High School/Mukilteo, Washington (Project Engineer)
Seattle Heights/Seattle, Washington (Project Engineer)
Harbor Steps/Seattle, Washington (Project Engineer)
Pacific Heights/Seattle, Washington (Project Engineer)
Seahurst Elementary School/Seattle, Washington (Project Engineer)
Seattle Tower/Seattle, Washington (Project Engineer)
Port of Seattle Pier 69 Headquarters Building/Seattle, Washington (Project Engineer)
Port of Seattle Central Waterfront Development, Seawater Cooling/Seattle, Washington (Project Engineer)

Commercial Site Improvements

Man Power, US Bank Center/Seattle, Washington (Project Engineer)
Microsoft Corporation/Houston, Texas (Project Engineer)
Seafirst Fitness Center, Seafirst/Seattle, Washington (Project Engineer)
Davis Wright Tremaine, Century Square/Seattle, Washington (Project Engineer)
HDR Engineering, Koll Center/Bellevue, Washington (Project Engineer)

Public/Government Facilities Projects

Chief Leschi School K-12/Puyallup Indian Reservation, Washington (Project Engineer)
VA Hospital Ventilation Improvements/American Lake, Washington (Project Engineer)
Cascade Middle School/Burien, Washington (Project Engineer)
Kamiak High School/Mukilteo, Washington (Project Engineer)
Mariner High School/Mukilteo, Washington (Project Engineer)
Seahurst Elementary School/Seattle, Washington (Project Engineer)

Residential Construction

Harbor Steps Stairs/Seattle, Washington (Project Engineer)
Harbor Steps Residences/Seattle Washington (Project Engineer)

If you have a long employment history, elaborate the relevant positions and use the short form for the others (years, company name and address, job title). Usually there is not too much from your work in the 1970s or 1980s that will help you get a position after the year 2000. Do not elaborate on old jobs. There are always exceptions, though. For example, one electronics technician had twelve years of experience in electronics. Then he sold automobiles for seven years because his brother-in-law owned a dealership. After that stint he found himself back in electronics. Just as in the case of a broken chronology in an education section, the solution was to construct a section called *Related Employment* and move all the earlier experience in electronics to the top of the page. The car sales came under *Other Employment,* which was cut short to get on with the rest of his material.

Military work history can be listed under either *employment* or *military.* If you have a service record, it is important. At the least, it will demonstrate administrative skills, since everyone achieves some rank that involves administrative duties. This is one angle that can be developed: "In charge of fourteen-man field team . . . ." More importantly, list both technical and administrative military programs you completed. Keep photocopies of the certificates in your folder for interviews. I find that servicemen and-women tend to underplay their service years, and they grumble about *black box technology.* Nonetheless, the experience is important in your employment profile and should be marketed.

If your employment record merits a few pages or more, then you should construct a separate document as a *supplement* to the resume. You probably want the resume to be one to three pages long. Two pages may be a practical goal. If you have a large block of material that will interrupt the flow of the resume, then remove most of it and construct a supplement. Teresa's list (Sample 11.L) *is* her supplement to her resume. Here she lists her engineering projects. Keep the best of the best in the resume, and place the rest in a separate document, indicated in the resume, as follows:

(For additional information please see the employment amplification.)

Engineers often find that the diversity of their work merits either an extensive explanation or an extensive list of projects. Teresa's supplement is a typical example. Although she is a very young civil engineer, she has had a wide range of experience, but the list was too awkward for the resume.

Sample 11.M

Resume of: 3188 Jensen Rd., Apt. 1
                                                    Seattle WA 98167

JAMES ELLIOT   (206) 365–0000

JOB OBJECTIVE

An entry level position that will allow me to concentrate on the field of electrical control while also providing the opportunity to expand my technical skills.

EDUCATION

- North Seattle Community College
9600 College Way N, Seattle, Washington

> _Graduation:_ June 8, 2000
> _Degree:_ Associate of Applied Science Degree in Electrical Power and Control
> _GPA:_ 4.00
> _Academic Honors:_ Dean's List Fall 1998 to Spring 2000
> _Major Areas of Study:_ Electrical theory (DC and AC); solid state fundamentals, advanced solid state, integrated circuits, digital logic, computer fundamentals, programming (basic and assembly), rotating machines (DC and AC), static control systems, industrial measurement and control, interfacing (computers to machinery), electrical code design, hydraulics
> _Support Classes:_ Physics, Chemistry, Pre-calculus, Electronics Drafting, Soldering and Hand Tools, Technical Report Writing

- Shoreline Community College
16101 Greenwood Ave. N, Seattle, Washington

> _Graduation:_ December 15, 1997
> _Degree:_ Associate Degree in Arts and Sciences
> _GPA:_ 3.34

EXPERIENCE

1995 to Present—Sorter, United Parcel Service, Seattle, Washington Received Employee of the Month honors three times since the program was started in 1997.
1994 (seasonal)—Landscaper, Pro Landscaping, Bellevue, Washington. Left job to continue my education

MACHINES AND EQUIPMENT USED

Oscilloscopes	DC static drives	AC motors
Magnetic starters	DC motors	Relays (current and voltage)
Servomotors	Curve tracers	Stepper motors
Transformers	Generators	6502 Microprocessor
Hydraulic motors	8080 Microprocessor	Servovalves
Programmable controllers		

REFERENCES

Provided upon request

Computer Know How

It is very apparent to me that the current generation of engineering and engineering technology students take their computer skills very much for granted. Many graduates have a tendency to overlook these skills when they package a resume. In truth, computer savvy is *very* valuable, and prospective employers will be delighted to see the extent of your familiarity with software and hardware. Consider including a list of software applications you know and the system hardware you have used. Include data processing languages also. This material can be identified under a major heading if you are a CAD specialist (see Sample 11.N), a program designer, or if you are directly involved with computer technology (see discussion on pp. 199 and 401). For other engineering specialties, your itemized computer skills could appear as a subsection under education or under employment. This is a rather recent issue, so there is no conventional place to put the information. Prioritize the list and put it where it seems most appropriate in terms of your technical specialty. (Similar lists of important equipment familiarity are common in resumes, as you will note in Sample 11.M).

Working knowledge of DOS 6.22, NT Workstation 3.51 and 4.0, Windows 3.11, 95, 98, UNIX, Novel 3.12.

Installation and configuration of all hardware and peripherals devices including laptops and radio frequency tablets and all types of local and networked printers.

Working knowledge of Microsoft Office 97 from end user perspective, including MS-mail, Exchange, Outlook 97, 98.

Excellent troubleshooting skills.

Knowledgeable in SMS (System Management Server).

Experienced in assembling CAT5 network cable and installing new network connections.

Exercises Chapter 11, Section C

Employment:

_____

Outline your employment history. Be thorough in identifying the most recent ten years of your job history and any other employment that is related to your interests or is otherwise important to you. Make the outline into a list, from your current position backward in reverse chronology. Make three columns: dates, employer (with an address and supervisor), and position held.

_____

Develop a paragraph description to explain the three most important positions.

_____

Keep this document in your portfolio in the event that an employer requests a more thorough employment profile than the one available on the resume. It serves as a practical reminder you can review prior to interviews. You can take selections from the document and use them in a cover letter for the resume or an application letter for employment (which are often combined into one letter).

_____

CHAPTER 11   THE RESUME: YOUR CORPORATE PASSPORT

Awards, Certifications, Interests, References

Personal Information

You are more than your job titles. You are more than your skills. Toward the end of the resume, you have the opportunity to say a few things about yourself. Again, the issue is marketing. If you have an old book on resume writing, it will include comical suggestions for including personal information such as the phrase *own home,* which at one time either carried some special status that has been forgotten or was meant to be translated as *need work.* In those days, people would also include their social security number. In this age of electronic highways, keep all such numbers to yourself. Other odd items of the time were observations on the order of *have driver's license* or *wife and three children.* Times change.

Nowadays, these considerations are ignored. They are not illegal, as some would believe. Considerable confusion has resulted from the civil rights acts. These acts have restricted what an employer can ask you, but they have little or no effect on what you can tell an employer. I would not eliminate the personal section, but I would certainly redefine its contents. There are a number of appropriate details that are quite valuable.

- *Certificates and licenses* If you are in a field that involves certifications, they should be identified. At times, this information may include unusual material, for example:

Specialty Licenses:

FAA Private Pilot Single Engine Sea and Land

FCC General Radiotelephone with Marine Radar Endorsement

Commercial Driver's License

- *Clearances* It costs thousands of dollars to clear an employee for classified work. If you had a *secret, top secret,* or *crypto-secret* clearance, be sure to mention it. This moves you up the ladder of applicants because you are saving a corporation money before the company even meets you. Personnel officers may have to clear you only from the date of your discharge. This qualification carries a lot of weight with military subcontractors.

Sample 11.N

Lincoln Harris   **CAD Drafting Technician**

5701 3rd Ave West • Seattle WA 981100 • (206) 491–0000

Careet Objective:
Electro-Mechanical, Industrial Drafter/Designer for an Engineering Manufacturing firm. Willing to take on any drafting-related function. Motivated towards responsibility and challenge.

RELATED EXPERIENCE:

1997–present ENGINEERING DRAFTER
KRAMER, CHIN, AND MAYO;
1917 –1st Ave., Seattle, Washington, 98101
Civil Mechanical and Structural Drafting Using CAD

1995–1997   ENGINEERING DRAFTER
THOMPSON ELECTRICAL CONTRACTORS, INC.
301 4th Ave, Seattle, WA 98106
Major responsibilities include drawing new projects, as-built drawings, preparing submittals, operation & maintenance manuals, etc. Assist estimators in quantity surveys and write-ups.

COMPUTER SOFTWARE EXPERIENCE:
— Microsoft Disk Operating System (MS-DOS) 7.0
— Automated Parts List Generator (APLG)
— Drawing Index On-line (CIOL)   —CC:Mail 2.21
— CADKEY Version 97   —AUTOCAD Releases 13 and 14
— WordPerfect 7.0   —Microsoft Word 7.0

EDUCATION:
NORTH SEATTLE COMMUNITY COLLEGE
9600 College Way North Seattle WA 98103
Graduated in 1995 with Associate of Applied Science Degree in Electro-Mechanical Drafting

ACHIEVEMENTS:
Recipient of the 1998 Outstanding Employee Award
Recipient of the 1994 Washington Award for Vocational Excellence scholarship.
Recipient of the 1994–1995 National Dean's List Certificates (National Honor Society).
College Dean's List (College Honor Society) 1994–1995.
Member of the Phi Theta Kappa, chapter Alpha Epsilon Omega.

REFERENCES AVAILABLE UPON REQUEST

- *Awards* Brag a little. Attendance awards. Employee of the month or year awards. Military honors. Lincoln focused his resume on computer software, but he did not overlook his employee awards (Sample 11.N).

- *Publications, patents* I recently had a student who has written articles for years for the national Saab collectors club newsletter. People have skills they often overlook.

- *Civic activities, volunteer services* The fire department captains I have worked with tell me that this is the first place they look on a resume. Volunteer fire or emergency medical service will help, but they want the team players. They are looking for the Little League coach or the scout headmaster or the YWCA counselor. (Avoid mentioning specific church affiliations in this category. It is a traditional taboo. Mention *church* but not the specific faith.)

- *Affiliations* Identify memberships in professional societies. Your memberships may help reveal the seriousness of your career interests, and the memberships will provide you with valuable periodicals and job announcements. There is usually an inexpensive student membership rate. Here is a sample entry:

AFFILIATIONS:

VICA (Vocational Industrial Clubs of America)

IEEE (Institute of Electrical and Electronics Engineers)

Lettermen's Club

Hispanic Engineering Society, Student Affiliate

- *Personal interests* You may have personal hobbies you want to mention. Computer enthusiasts, for example, are often involved in related hobbies. Do *not* mention high-risk sports. If you mountain climb or if you built your own ultralight, you might just say you enjoy chess instead. Companies want to pay your tuition for continuing education; they do not want to send you flowers.

- *Health* The expressions *Nonsmoker* or *Nonsmoker, nondrinker* are comments that will be respected in the age of corporate urinalysis tests. This is an older convention but such comments reflect upon the health of the applicant, and corporations are probably inclined to take the observations seriously.

- *Disabilities* If you are disabled in any way, feel free to mention the status in some reasonably specific way. Some states have a numbering system—D-4—and some agencies have other measures such as the amount of weight you can be expected to lift: *25 pounds maximum.* You can also simply identify the injury. Do not feel

Sample 11.0

<u>Elizabeth Anderson</u>
15455 Pine St.
Seattle WA 98122

Phone: (206) 367-0000

JOB OBJECTIVE

To locate a position in the field of Electronics Technology as an Engineering Aide or Technician. Most interested in design and development support.

EDUCATION

1998–Present   North Seattle Community College
     9600 College Way North
     Seattle WA 98103

     Specialized courses include:
         Advanced Solid State Circuits, Digital Fundamentals
         and Applications, Microcomputer Fundamentals, Basic
         Programming, 6502-based Assembly Language
         Programming, Calculus, Technical Report Writing.

Graduation: December 2000 with Associate of Applied Science Degree
     Digital Electronics Technology.

1997–1998   ITT School of Business and Technology
     110 West Smithton
     Seattle WA 98310
       Completed one year of electronics studies covering basic electrical theory, transistor theory and applications, and communications electronics.

EMPLOYMENT

1999-Present   KOMO News/Fisher Broadcasting Inc.
     4th North & Denny Way
     Seattle WA 98156

Six-month externship in cooperative work experience designed to provide students with experience in communications systems maintenance.

1997–1999   Etron
     12629 Interurban Avenue South
     Seattle WA 98168

<u>Customer Service Assistant</u>
Assistant to repair technicians involving receipt of faulty equipment, inspection, calibration, testing and repair of microwave door openers and security devices.

REFERENCES AVAILABLE ON REQUEST

awkward about this matter. I deal with a number of rehabilitation situations each year, and these men and women learn to realize that corporations consider them to be serious and dedicated workers *because* of the injuries. Besides, disabled workers fall under affirmative action guidelines, and industry looks for talent in these employment pools. If you list a disability, be sure to also list *health* as a category and say *excellent*. Many healthy people are disabled; you want to make this point.

Since there is considerable misunderstanding about this small section of the resume, you can cut an edge here. Because many people have the wrong impression, they omit the personal section at their own expense. The old style of personal information was hardly confidential, but it was certainly not a marketable product. It is just as well that the information is forgotten. However, you do not want to miss any opportunity to share relevant information that matters. You can insert whatever is of interest. You can even upgrade the section title by eliminating the term *Personal* and calling it *Awards, Qualifications, and Interests* or whatever fits your needs.

References

By tradition, references go last. It is hoped that, this is proof that you will get the job based on *what you know* and not on *whom you know*. By convention, it is also perfectly appropriate to omit the names. Do *not,* however, omit the reference section. It should always say:

> References available upon request.

You will need to take the comment seriously. Since Elizabeth (Sample 11.O) used the *available upon request* strategy, she must have the references ready and waiting.

You may list your references (perhaps two, three, or four) on a separate sheet of paper. This should be ready at the time of an interview so that you can then hand a potential employer the list of names, with addresses and phone numbers. You may also list your references on the resume. For many college graduates, including references in the body of the resume is a very mechanical decision. If they design nice resumes that stop at the bottom of a page or two, they will omit the list, since it starts another page. However, if a page is half full, why waste the half that is empty? Include the references. There are no rules, as I say, although some resume workshops will insist that you do not include references. Your references are a strong link between you and your career field. I definitely like to see them, and I like to see them handled correctly.

The references you would list on a resume are what are called *professional references*. They are not character references of the sort that a bank, for example, will request for a loan. Professional references concern your knowledge and your skills. They can be a mix of academic references and former employers. You usually list several former supervisors from work. If you have no former employers, you might list several college

Sample 11.P

Earl Wasaki, Instructor
Electronics Technology Program
Engineering Technologies Division
North Seattle Community College
9600 College Way North Seattle WA 98103
Phone (206) 634-3100

Paul C. Davis, Instructor in Electronics
Technical Programs Department
Mountlake Terrace Technical College
Mountlake Terrace WA 98043
Phone (206) 778-8921

William Steen, Test Supervisor
Eldec Corporation
19230 Highway 99 Lynnwood WA 98036
Phone (206) 771-7216

instructors, preferably in your technical specialty. You do *not* list friends or family. Do not list anyone you have not consulted. If you have permission to include a reference, but the permission dates back several years, there could be a problem. You should make sure the address and phone numbers are correct and ask for a new opportunity to use the person as a reference. At the least, such updates remind your references of your situation in case they are contacted.

A second point concerns the manner of presentation. As I will explain at the end of this chapter, the resume is somewhat of a contradiction: it is very short, but it is very thorough. Do not simply identify a name and phone number.

George Benini

321-4078

You must market the product, so be sure the reader sees the importance of the reference.

(Name, Title)	**Dr. George Benini, Instructor**
(Department)	**Structural Engineering Program**
(Division)	**Engineering Technologies Division**
(Institution)	**Seattle Community College**
(Address)	**9600 College Way North Seattle, Washington 98105**
(Phone)	**(206) 321-0000**

If you are thorough, the reference section will speak well of you by identifying professionals who respect your work and who have shared a working environment or academic environment with you. Notice the references in Sample 11.P. This can be a separate document or part of the resume. Either way, it reflects valuable professional associations.

Exercises Chapter 11, Section D

Portfolio:

Gather the appropriate documents that will support your interviews and prepare copies of them to carry to interviews. These items will include documents of interest such as the following:

- Degrees
- Certifications (including military)
- Awards (employee of the month, and so on)
- Transcript of grades
- Honors
- Specialty licenses
- List of references
- Other

For a list of references, identify three former or current supervisors or college instructors who are familiar with your skills and knowledge. These are not "personal" references. Identify the person, the title, the agency, the company address, and the phone number. (At this point you do not have to ask permission for the references, since these are preparation steps.)

Final Considerations

If you return to the list of popularly held ideas about resumes (see p. 375), you will realize that I do not agree with any of them. The *Wall Street Journal* conducted an extensive poll some years ago regarding the commonly-held idea that resumes have to fit on one sheet of paper. As anyone could guess, most respondents said there had better be more to a prospective employee than can fit on one sheet of paper.

A few additional tips may be of value and this discussion will involve yet another conflict about how to build a resume. In recent years the device known as an *action verb* has become increasingly popular. Many of the resume workshops and resume self-help books encourage the use of action verbs. This tactic means that you are supposed to take verbs very seriously and give them special attention. You may even be told to boldface them. A typical list identifies them for you. Here is the beginning of an alphabetical sample of such a list:

ACTION VERBS		
accomplished	completed	designed
achieved	conceived	determined
administered	concluded	developed
advised	conducted	devised
analyzed	continued	directed
applied	contracted	drafted
appraised	controlled	edited
arranged	coordinated	ensured
assisted	corrected	established
assured	counseled	estimated
briefed	critiqued	evaluated
bought	dealt	expanded
budgeted	decided	expedited
cataloged	defined	explained
changed	delegated	forecast
classified	delivered	formulated
compared	demonstrated	⋮

Sample 11.Q

<u>ANNA RITTER</u>

404 S. Westfield St.
Seattle Washington 98118
(206) 935-0000

JOB OBJECTIVE

Seeking a technical support position in the electrical power and control field. Emphasis on the interfacing and application of computers in industrial control.

EDUCATION

Associate of Applied Science Degree in Electrical Power and Control Technology. North Seattle Community College, 9600 College Way North, Seattle, Washington September 1998 to June 2000.

<u>Program Content</u> (main courses)

Electrical Theory AC/DC
Solid State Fundamentals
Integrated Circuits and Advanced Solid State
Digital Logic and Computer Fundamentals
AC and DC Machines
Instrumentation and Servo Systems (Hydraulic & Electrical)
Static Control Systems and Electrical Code Design
Technical Math for Electronics
Technical Physics and Technical Drafting

EDUCATION HIGHLIGHTS

Three-phase power distribution systems, dynamos (all types), programmable controllers and interfacing, electrical controllers, stepping motors, static drivers (AC and DC).

Practical applications of digital and analog circuitry.

All subjects included hands-on experience with the equipment.

Designed and built a hydraulic robotics milling machine utilizing digital gates, analog circuits, and servo-valves to control hydraulic actuators.

REFERENCES

Available upon request.

I seriously question the utility of such a tactic. It is harmless, of course, but the question is, will it achieve any mission in particular? I am skeptical that it will—at least if you are an engineer. If you look carefully, you will note that by far the lion's share of the list consists of *managerial* verbs, not action verbs. In fact, I would guess that whoever first proposed this tactic had managers—probably MBAs—in mind. The problem is that this practice is now widespread. A college graduate is in no position to use managerial verbs, and I would argue that it is not an ideal tactic for accomplished engineers either.

The truth is that the verbs are empty. Unless you are a manager, the verbs do not address skills, knowledge, and hardware. Action verbs are for power players. Instead, I would urge you to use a different tactic. Use nouns. Be specific. Identify programs you have completed, tools you are familiar with, or computer applications you use regularly. Similarly, be sure to add addresses of former employers and titles for former supervisors. Obviously, you do this to be thorough, but from a marketing point of view, the sheer power of nouns and proper nouns is a psychological and strategic matter.

Look at Sharon's education and employment details for an example (p. 364). Nouns impress. The concrete image is critical. Go a step further by adding such *things* as course work or lab equipment you are familiar with, or identify fieldwork projects you completed. There are almost no verbs on the entire page of Teresa's very strong profossional project history (p. 384). Review Mike's resume (Sample 11.G) and notice the intensity of the nouns. This resume was very impressive, and I happen to know that the employer who hired him was persuaded to do so by the resume. Anna's resume on p. 398 also was designed with nouns in mind (Sample 11.Q). Notice the course work detailed in Michael's resume (Sample 11.R).

One reason that a noun emphasis can be quite powerful is because it can be readily constructed into the first half-page of the resume, where it can be used to identify a list of employment qualifications. Here, for example, is the opening of a resume of a CAD drafter:

Capabilities:

- Electronics drafting involving schematic diagrams and circuit board layout using PADS per ANSI Y32.2 and ANSI Y32.14.
- Geometric Tolerancing per ANSI Y14.5M-1982 and ASME Y14.5M-1994.
- AutoCAD R14 Advanced Applications, 2D and Solid Modeling.
- CADKEY 6.0 Fundamentals with 3D Applications.
- Orthographic, isometric, assembly, and detail drawings along with BOM compilation.
- Experience with sheet-metal design and fabrication, including flat-pattern drawings.
- Manual drafting skills including lettering and line work.
- Proficient in DOS and Windows (Word, Excel, Access).

Sample 11.R

Michael Brandel

9553 Redwood Dr.
Redmond Washington 98036

Home Phone
(206) 774-3394

PROFESSIONAL OBJECTIVE

To obtain a position as a mechanical draftsperson, with the possibility of moving into engineering design.

EDUCATIONAL BACKGROUND

Associate of Applied Science Degree, Electromechanical Drafting
June 8, 1999
North Seattle Community College Seattle, Washington

General Studies
Sept. 1996–June 1997
Montana State University
Bozeman, Montana

RELEVANT COURSES

Industrial Drafting I–IV:

Orthographic drawings, Auxiliary drawings, Sectional drawings, Design drawings, Detail and Assembly drawings, Isometric and Oblique drawings, Structural drawings, Piping drawings (single and double line), Gear and Cam drawings, Welding and Riveting drawings, Thread and Fastener drawings

Manufacturing Processes:

Technical Mathematics (logarithms, trigonometry, tapers, cutting speeds and feeds, gears, threads, strength of materials, etc.)

Thermodynamics

Basic Computer Programming (HP 3000, Mac SE)
Technical Report Writing

PRIOR WORK EXPERIENCE

March–August 1999	Denny Manufacturing, Seattle, Washington Draftsman: Ink on mylar, detail and assembly drawings, customer drawings.
August 1997–August 1998	Gasket Engineering, Daytona, California Draftsman: Vellum and lead, all shop drawings from detail to customer drawings.
July 1990–July 1996	United States Navy Gunner's Mate Technician: maintenance and operation of antisubmarine rocket launcher Top Secret security clearance
REFERENCES	Upon request

The following sample, as you will realize, is not that of a young college graduate. This computer network specialist had years of experience in both Europe and the United States. In general, the use of a list of *qualifications* depends on at least a few years of professional practice in the field.

- Knowledge of MS-DOS, Windows for WorkGroups 3.11, Windows 95, Windows 98, Windows NT, and NetWare operating systems

- Experience in Novell and Windows NT LAN/WAN network administration

- Ability to install, configure, optimize, troubleshoot, and maintain computer and network system hardware and software

- Seattle Community Colleges fiber optic wide area network project manager

- Seattle Community Colleges Internet II (Gigapop) project management team member

- Knowledge of HTML, Qbasic, and Visual Basic programming; knowledge of Internet and browser programs such as MS Internet Explorer and Netscape Navigator (Communicator); knowledge of communications and e-mail programs such as Shiva/Dial-Up Networking, Minisoft 92, Telnet, FTP, Exchange, Outlook, GroupWise, CC:Mail

- Knowledge of MS Office 95/97/2000 and Corel WordPerfect Suites 7/8/2000

- Knowledge of multimedia presentation programs such as Authorware, MS PowerPoint, Astound, and Corel Presentation; technical programs such as Visio 5, MS Project 98, 2D and 3D Designer

- Knowledge of graphics, publishing, Web page creation, and editor programs such as CorelDraw, PhotoShop, Corel PhotoPaint 8, MS Publisher, PagePlus 5.0, HotDog Pro, Front Page, and others

- Hewlett-Packard NetServer Technical Training and Mannesmann Tally laser printer factory trained

All the resume samples you see here use the technical-noun strategy rather than the action-verb strategy. You will find only one exception if you review them. See what you think. You be the judge.

Basically, you want the document to be as hard and factual as you can make it. In manuals for writers, this is, with no pun intended, called the concrete style. Use nouns and avoid abbreviations. There is a tendency for a writer to clip the resume to make it shorter, but this should not be done at the expense of image-building. Your first concern is your image; it must be your priority. It would be nice if the resume could be a quick and easy read, but this is not a priority *if* you effectively interest the reader on the top of the first page. Besides, if you write out whatever abbreviations you use, you will probably add little more than an inch to the resume. Your image is more important than economy of space.

Sample 11.S

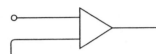

Katherine Galligan
1040 Williams Mill Rd.
Bellevue WA 98007
(206) 641-0000

OBJECTIVE: Electronics Technician in design of circuits.

EDUCATION:

1994–1999:
North Seattle Community College
9600 College Way N
Seattle WA 98103
Degrees: –Associate of Science in Electronic Communications
–Associate of Applied Sciences in Digital Electronics
–Associate of Applied Sciences in Electrical Engineering
Technologies

1993–1994:
Lake Washington Vocational Technical Institute
11605 132nd Avenue NE
Kirkland WA 98033
Area of study: Electronics

1992–1993:
Bellevue Community College
3000 Landerholm Circle SE
Bellevue WA 98007
Area of study: Business

EXPERIENCE:

1995–Present:
Cascade Geophysics   Supervisor: Bill O'Connor
1121 Bothell Way
Bothell WA 98001
Position: Administrative Assistant
Backup tapes, report documentation, client assistance,
company function arrangements, purchasing

1990–1995:
Albertson's   Supervisor: Richard Hansen
403 Bellevue Way
Bellevue WA 98005
Position: Manager

1987–1990:
REI   Supervisor: John Falcone
155 NE 24
Bellevue WA 98007
Position: Clerk

REFERENCES:
Available upon request.

The point is that the resume can become too dense if it is condensed in either layout or content. If you jam the page, the resume looks congested and will not project clarity. Similarly, you want to be as concrete as you can be, and I personally think you do not want to spend years of hard work and then say you have an

A.A.S.

You want to stand up and clear your throat and say,

Associate of Applied Science Degree in Heating, Ventilation, and Air-Conditioning Technology

The difference is obvious, so I discourage abbreviations, especially if you have three degrees, as in Katherine's case (Sample 11.S). If you prefer, you could even eliminate the abbreviations of the states identified in your personal address or business addresses, although frequent repetition of *Pennsylvania*, for example, might not be desirable. Make it concrete. The reading time is not affected; in fact, I would argue that an excess of abbreviations will slow a reader down!

Some abbreviations may still seem to you to be appropriate. Perhaps the original or complete version is long, uncommon, or irksome. Perhaps there is a repetition problem and the long version will become clumsy. In such cases, consider using the long version for the first occurrence, and use the abbreviation thereafter. Strike a balance, emphasizing image first, brevity second.

As you can see, everyone designs his or her own idea of what a resume should be. The lore is conventional, the styles are traditional, but, with a few special touches, every resume can be original. You looked at Robert's resume earlier (Sample 11.K). The version on Sample 11.T was the alternative model designed with tables! This is very innovative. Katherine's resume (Sample 11.S) used a border with an electronics symbol. She said the device attracted interest. You could use your own company logo on resumes that are used to promote your business. Several of the resume samples contained elements in color.

In the end, you will decide for yourself on all matters regarding your resume—but *do* have someone review it for readability. What may seem very clear to you may not be clear to a reader. You will have pored over the resume for hours. You need readers who have never seen it to test it. Simply note any questions they have and get right back to the drawing board. There should be no questions. If all goes well, there should be a job offer. Good luck.

Sample 11.T

Robert Kangas

102 Main Street, Apt. F
Northland WA 98011-3402
(206) 489-0000

OBJECTIVE

To obtain full-time, permanent work where I can best utilize my skills in integration and installation of PCs, troubleshooting, and repair.

EXPERIENCE

1997 to present	Design, build and install control systems using Pentium-based computer systems. Technical Support at Phototronics Corporation, 5110 190th Street NW, Suite 100, Lynnwood, WA 98036 (206) 776-0000.
1993–1995	Integrate and install IBM AT, clones for use with DOS and UNIX office systems, including cabling of terminals and printers. Repair computers, terminals, tape drives, modems and printers. Technical Support at DXD Computer Systems, Lynnwood, Washington.
1990–1993	Repair, maintain, and install CPT Word processors, printers, and scanners. Field Service Technician at Proline, INC., Seattle, Washington.
1986–1990	Customer Service Representative at Singer Business Machines/WRT Customer Service Division, Bellevue, Washington.

EDUCATION

North Seattle Community College 9600 College Way North Seattle WA 98103 (206) 527-3600	(April 1999) Associate of Applied Science Degree in Electronics Technology. Solid State, Digital, and Analog Electronics. Troubleshooting, Industrial Applications. Discrete Circuit Analysis & Application, General Physics, Calculus, Technical Writing.
Singer Business Machines/WRT	Company schools: Point-of-Sale Terminals and Docutle Automatic Teller Machines.
U. S. Air Force 1981–1985	Electronics, Digital electronics, and Computer maintenance.
Meadow High School Renton, Washington	Electronics and Computer math.

INTERESTS

Amateur Radio   KB8WNK

References provided upon request.

Exercises Chapter 11, Section E

Resume Development:

Build a resume that targets a potential employer and focus on a company that would offer a typical job opportunity in your area of specialization. The resume can be current, projected, or theoretical. In other words, the resume can be designed to reflect your current status; it can be designed to reflect work completed upon your graduation (let's say in six months); or it can be a conjecture based on further studies at a four-year institution. In each case, list course work that you have completed or will complete.

This is an exercise in design and accuracy. Build the resume with the intention of using it. Be thorough and thoughtful. (The four-year degree resume is the exception, of course. If you build a resume with a distant graduation in mind you obviously cannot be thorough and you cannot use it in the near future, but you will have the design constructed and ready to go.)

Use the following guidelines to build the document:

- An objective
- Special skills or qualifications
- Education
- Employment
- Military history
- Computer know-how
- Awards and certifications
- Personal interests
- References

Citation and Documenta-tion Practices

This appendix selectively repeats the contents of Chapter 9, *Cues and Keys*. There are a number of popular systems that can be used to develop source citations and references. MLA conventions are identified in Chapter 9. MLA conventions are the standards adopted by the Modern Languages Association. If your instructor prefers that you use APA conventions, use this appendix as your guideline for citation and reference practices. APA conventions are those standards practiced by the American Psychological Association.

This appendix does not repeat the general discussion concerning the uses of citations and references. For an appropriate understanding of documentation practices, you should first read Chapter 9. Then, if your instructor prefers APA conventions, read this appendix. The distinctions between the MLA practices and the APA practices may appear to you to be subtle matters. For this reason, citations and references are easy to understand since the various systems are similar. If, however, you are asked to develop your own citations or references for a document, you will need to know the *precise* methods for a specific system. If MLA standards are preferred, follow the samples illustrated in Chapter 9. If APA standards are preferred, follow the samples illustrated in this appendix.

 Note: There is a software application for APA format.
Format Ease, Version 2.0
Guilford Publications, Inc.
72 Spring St.
New York, New York 10012-4019
Phone: 212-431-9800
http://www.FormatEase.com

Source Citations

If you borrow information from another source—a book for example—you must acknowledge the source. This information might be a quotation, but in engineering and technical areas the information is more likely to be paraphrased, in which case you explain, in your own terms, material from the other source. Because the paraphrased material is conceptually the same as the original material, it is considered "borrowed," and the original source must be acknowledged. Statistical material or research findings also must be acknowledged if the source of the material is not that of the author.

In APA format, citations are briefly handled in the text as you will note in the following sample.

One Work By One Author

Full parenthetical citation

The parsec is a unit of distance in astronomy and not a unit of time, as suggested in the first *StarWars* trilogy **(Downing, 1997).**

The citation format is based on the following prototype:

Borrowed information **(author, year of publication)**

The citation is enclosed in parentheses and the author's name is followed by a comma. Note that if the parenthetical citation occurs at the end of a sentence, the period is placed after the second parenthesis.

If the name of the author is a part of the running text of the sentence (sometimes called a *tag*), only the date is used to indicate a citation.

One Work By One Author

Author tag citation

According to **Barrett (1993),** black holes are formed as a result of the death of very large stars, and the compressed matter that remains has a gravitational pull that is strong enough to keep radiation (such as light) from escaping.

If you construct a sentence that refers to both the author and the year of publication, there is no reason to include a parenthetical citation.

One Work By One Author

Narrative citation

Lemay's findings in 1994 confirmed that REM sleep occurs in four to five periods of eye activity during sleep.

If there is more than one author, cite all authors by last name and the date of publication. Do not alphabetize the authors. Present the list of authors as provided on the title page of the document.

One Work By Multiple Authors

Full parenthetical citation

Electrons do not follow the neat circular paths depicted in textbooks **(Wiley, Bevin, and Mitchell, 1981).**

This citation includes the word *and.* Because multiple-author citations are awkward, subsequent citations are abbreviated by using "et al." after the first author's name.

One Work By Multiple Authors

Subsequent parenthetical citation

The term *magic bullet* was coined by the German physician Paul Ehrlich and refers to a new avenue of drug research that is designed to create drugs that will target specific diseased tissues **(Talmidge et al., 1991).**

Note that in the preceding sample *et al.* (from the Latin *et alii,* "and others") is not underlined or in italics, and note that al. is followed by a period.

Several other variations of the citation format are common in technical documents. If your source is the product of an agency, a society, an association, or a similar group, the group is acknowledged as a collective author.

A Group as Author

Full parenthetical citation

The mission photographs revealed that some of the moons of Jupiter are volcanically active **(National Aeronautics and Space Administration [NASA], 1987).**

The abbreviation of the group is identified in the citation within brackets, so that it can be used in subsequent citations.

A wide variety of brochures, manuals, pamphlets, and other documents are produced by corporations. These documents will seldom identify an author. They are identified by title, or a shortened version of a long title.

A Work With No Author

Full parenthetical citation

The handheld Magellan uses the Global Positioning System (GPS) and is accurate anywhere in any weather **(Introducing the Magellan GPS Meridian, 1993).**

Underline the title of a corporate document in which the author is not identified. In the preceding example, a subsequent citation could be shortened to *Magellan GPS Meridian* or *GPS Meridian.* or perhaps, *Meridian.* The shortened form is a convenience, but only if the reference to the original document remains clear to your reader.

The final model illustrates a citation in which a number of resources are identified. Because research regarding any particular investigation may be conducted by many scientists, it is quite common to see a number of sources acknowledged. In this case, a writer is often not paraphrasing but referring to a body of literature or findings.

Two or More Works By Different Authors

Full parenthetical citation

Other investigators were less confident that the shale indicated significant oil deposits **(Harvey, 1981; Sanders, 1995; Tobin, 1990).**

Observe that the authors are alphabetical and the citations are separated by semicolons. The word *and* is not placed before the final name.

You should be able to structure most of your citations by following the samples in the discussion. If you happen to use a quotation, two general rules will be helpful

1)   Quotes of fewer than 40 words are placed in the running text and are enclosed in quotation marks.

2)   Longer quotes are set off from the text and intended five spaces to seven spaces (on the left side only) and double-spaced without quotation marks.

A Note Concerning Publication

Note that APA standards are primarily designed for manuscripts that have not been published. Published quotes are usually single-spaced, which is a widely accepted practice. If your project conforms to APA manuscript practices, double-space a long quote. If you are in a business setting that publishes its own materials, follow the prescribed style.

APA References

Once you have constructed a text using source citations, you must then list all the citations alphabetically at the end of the document on a numbered page with the title *Reference List* at the top center. Identify all the resources in the APA variation of the familiar pattern that is used for bibliographies and reference lists. Double-space each entry and structure it in the format of a small paragraph in which the first line is indented, as follows:

```
          Doe.  J.  (Date).
        Title. City: Publisher.
```

The specific rules governing the generic model can be explained by creating a slightly more detailed model.

```
                    Entry for a Book

    Last Name, First and Middle Initials. (Date of Publication). Book title with

    only the first word capitalized. City: Publisher.
```

There are three unique features to the APA system:

1) Entries are indented in double-space format.

2) Full first and middle names are omitted.

3) Only the first word of a title is capitalized.

The following sample follows the APA format:

```
        Downing, D. A. & Covington, M. A. (1998)

    Dictionary of computer and internet terms. New York: Barron's Educational

    Series.
```

The details concerning this entry are as follows:

1) **Author.** Last name first—comma—first and middle initials with periods.

2) **Date.** In parentheses—use the most recent date in the copy you are using—followed by a period.

3) **Title.** Full title of book with only the first word capitalized—underlined*—followed by a period.

4) **City.** City where publisher is located—colon. Add state (U.S. Postal Service abbreviations) or country name if city is not well known.

With the state included, the order is city—comma— state abbreviation or country—colon

5) **Publisher.** The publisher—period. Such words as *company* or *corp.* are often omitted.

You can use italics for titles if the document is not a manuscript to be submitted for publication.

The most likely types of references that you will encounter in your work in a technical specialty include the following.

1) **A book**

2) **A magazine article**

3) **An article in a journal**

4) **An article in an anthology**

5) **An electronic source**

6) **A corporate document**

The following samples can be used as guidelines for the basic six type of resources:

Book

Name. (Date). <u>Title.</u> City: Publisher.

Sample Application:

Vaughan, T. (1998). <u>Multimedia: making it work.</u> Berkeley, CA: McGraw Hill.

The preceding entry is a reference to an entire text. If you want to indicate specific page references, add them after the publisher's name.

The time it takes to write and publish a book seriously limits the effectiveness of this type of resource if you are involved in disciplines that experience rapid changes. Computer-related literature is difficult to keep current in books. In areas such as network technology, the appropriate magazines and professional journals will often prove to be more up to date. As a result, these resources are more likely to be used as references. In a citation for a periodical you must include the article title *and* the periodical title.

Magazine Article

Name. (Date). Article. <u>Magazine, Volume Number,</u> Pages

Sample Application:

Pappalardo, D., & Gittlen, S. (1999, April). Users wait to sort out monitoring tools. <u>Network World, 16,</u> 10.

The month (and week date if appropriate) follow the year. Note that only the first word of the article title is capitalized, but the magazine title is conventionally capitalized. The volume number is underlined as part of the title. The page reference can indicate a specific page or the entire article.

The journals of professional societies are timely sources of technical and engineering information. The format is the same format used for trade magazines.

Journal Article

Name. (Date). Article. <u>Journal, Volume number,</u> Pages.

Sample Application:

Raya, S. P., & Udupa, J. K. (1990, March). Shape-based interpolation of multidimensional objects. <u>IEEE Transactions on Medical Imaging, 9,</u> 32–42.

Anthologies of articles can be useful sources of timely material. The anthology takes less time to create than a conventional book. Technical, scientific, and engineering organizations and publishing groups such as *Scientific American* publish anthologies.

Essay in an Anthology

Name. (Date). Essay. In Editor's name (Ed.), <u>Book</u> (Pages). City: Publisher.

Sample Application:

Parker, E. N. (1981). The solar wind. In J. C. Brandt (Ed.), <u>Comets: Readings from Scientific American</u> (pp. 76–84). San Francisco: W. H. Freeman.

Observe that an essay's author is used to identify the reference; however, the article cannot be located for research purposes without the name of the editor of the book. The unique feature used in this sort of entry is the word *In,* which signals that the editor's name and the book title will follow. Note that the editor is identified as such after his or her name: (*Ed.*) Note that the name appears with the surname last. Finally, observe that this particular reference uses the abbreviation for page or pages: *p.* or *pp.*

Electronic sources have become very popular and are frequently cited in lists of references. A CD-ROM is easily structured into a reference.

Electronic Source (CD-ROM)

Manufacturer. (Date). <u>Title</u> [CD-ROM]. Location.

Sample Application:

Microsoft Press. (1997). <u>Networking Essentials</u> [CD-ROM]. Remond, WA.

Another timely source of information is the Internet. If you use electronic sources, the conventions for citations and references are much the same as for conventional sources.

> ## Electronic Source (On-Line)
>
> **Author. (Date). Title. <u>Periodical or database</u> [On-line], <u>Volume number</u>. Available: Internet**
>
> **Address**
>
> Sample Application:
>
> Quint, V. (1998, October 22). The user interface domain. <u>World Wide Web Consortium</u> [On-line]. Available: http://www.w3.org/

Note that there is no period at the end of an on-line reference. Often, pieces of information are missing from an on-line reference. If the author is unidentified, place the title first, followed by the date.

> High speed and dedicated access (1998, November). <u>Dial-up</u> [On-line]. Available:
>
> http://www.seanet.com/

The reference format for an electronic source can also include the access date (the date you consulted the document). The formats of the two preceding samples conform to the 1994 APA style guide and do not use the access dates. If you want to include the access date, consult the APA's Web page updates (available at http://www.apa.org/journals/webref.html). Your instructor will suggest which format is appropriate.

The final sample reference that is of particular interest in engineering technical areas is the reference to a corporate document. Note that the first sample is a citation for an electronic source with an access date added in the newer APA reference style. Also note that most corporate literature does not identify authors. In APA format, the preferred reference technique uses a *corporate author,* meaning that the name of the company is used as the author.

> ## An Electronic Corporate Document with Access Date
>
> **Corporation. (Date). <u>Title</u> [On-line]. Retrieved [Date] from the World Wide Web: Internet**
>
> **Address.**
>
> Sample application:
>
> Adobe Systems Inc. (1998). <u>Preparing graphics for the world wide web</u> [On-line], Retrieved
>
> June 21, 2000, from the World Wide Web: http://www.adobe.com/studio/tipstechniques/GIFJPG
>
> chart/main.html/

You can use either technique for constructing an entry for a Website. The list of references on p. 297 and the list of references on the following page reflect the two methods. Use only one method for a given list of references.

Sample A.1

References

Berners-Lee, T. (1998). About the world wide web consortium. <u>World Wide Web Consortium</u> [On-line]. Retrieved November 15, 1999, from the World Wide Web: http://www.w3.org/

Cailliau, R. (1995). A little history of the world wide web. <u>World Wide Web Consortium</u> [On-line]. Retrieved November 10, 1999, from the World Wide Web: http://www.w3.org/

Castro, E. (1998). <u>HTML 4 for the world wide web: visual quickstart guide.</u> Berkeley, CA: Peachpit Press.

Cravotta, Nicholas. (1998, June). 21st century delivers quality service with java. <u>Networking Magazine,</u> 13, 70–74.

Lee, Lydia (1996, June 15). Java takes hold. <u>New Media,</u> 6, 46–54.

Lemay, L. (1997). <u>Teach yourself web publishing with HTML 4 in a week.</u> Indianapolis, IN: Sams.net Publishing.

Raggett, D., & Le Hors, A. (1998). Hypertext markup language. <u>World Wide Web Consortium</u> [On-line]. Retrieved November 14, 1999, from the World Wide Web:http://www.w3.org/

St. Laurent, S. (1998). <u>XML: a primer.</u> Foster City, CA: MIS:Press.

Spangler, Todd. (1998, May 4). Distributing web content to even out traffic loads. <u>Internet Worlds,</u> 4, 28–29.

Von Hagen, B. (1997). <u>SGML for dummies: a reference for the rest of us!.</u> Foster City, CA: IDG Books Worldwide.

References for conventional corporate publications—brochures manuals, pamphlets—follow the format of a book reference except that the corporation serves as the author.

> Adobe Systems Inc. (1994). <u>Adobe photoshop classroom in a book</u>. Indianapolis, IN: Macmillan Computer Publishing.

Sample A.1 is a list of references in APA format. The sample shows entries that are intended for APA manuscripts. If you are submitting a manuscript to an instructor or supervisor for editorial comment, the document is a manuscript and can use the indented format. If you revise the manuscript as "final copy" or plan on self-publishing, you can use the "hanging indent" style that is characteristic of reference lists (see p. 314).

Explanatory Notes

In APA manuscripts, explanatory notes are indicated by using superscripts in the traditional manner. The notes are placed on a separate page titled *Footnotes*. The numbered notes are double-spaced and indented in the manner of paragraphs. The following sample illustrates the use of superscript numbers.

> Coaxial cable (Figure 1), which has transmission rates of 10 mbps (millions of bits per second), was widely used as a backbone[2] for LANs prior to the invention of fiber-optic cable. However, it is still popular because it is inexpensive, lightweight, flexible, and easy to work with. It is also more resistant to interference and attenuation[3] than twisted-pair cable. Coaxial cable comes in two forms: thin (thinnet) and thick (thicknet).

On the page titled *Footnotes* the author included the two explanatory notes concerning this paragraph.

> [2] The trunk cable of the network.

> [3] Loss of signal strength as the signal travels greater distances.

Sample A.2 is a page from a paper concerning programmable logic controls that includes three of the six explanatory notes that were added by the author. In APA manuscript style, the notes were placed on a separate page such as appears in Sample A.3. Remember that these samples are *manuscript* practices. Many publishers place the notes at the foot (footnotes) of the appropriate pages once the manuscript is published. Other publishers will leave the notes grouped at the end (endnotes) of the published document, but single-spaced, usually with a double space between each single-spaced note.*

*The practices that are explained in this Appendix should represent most of the applications of concern to you. The Publications Manual of the American Psychological Association is a 350-page style guide from which I have selected only a few points of interest that may be of use if you want to develop citations and references. If you are interested in the full text, it is available at most university bookstores.

Sample A.2

Introduction to Programmable Logic Controllers

A Programmable Logic Controller (PLC) is a solid state[1] device designed to perform logic functions accomplished by electromechanical relays[2] (see Figure 1). PLCs were invented to replace the relay circuits for machine controls. The PLC works by looking at its inputs and, depending upon their state, turning on/off its output. The user enters a program, usually through software, that gives the desired results. PLCs are used in many "real world" applications where control of manufacturing process equipment and machinery is desired. The PLC is essentially computer designed instrumentation for use in machine control. It operates in the industrial environment and is equipped with special input/output interfaces and a control programming language.

Memory Size

There are variations in the size classifications of PLCs. They are divided into three major categories: small (also known as *micro*), medium, and large, each with distinct operating features. The small-size category covers units with up to 128 I/Os (input/ output) and memories of up to 2K bytes,[3] which are capable of providing simple to advanced levels of machine control. Medium PLCs have up to 2948 I/Os and memories of up to 32K bytes. They can be adapted to temperature, pressure, flow, weight, position, or any type of analog function commonly encountered in process control applications. Large PLCs are the most sophisticated units of the PLC family. They have up to 15,000 I/Os and memories of up to 2M bytes. They can control entire processing plants and have almost unlimited applications.

Sample A.3

4

Footnotes

[1] Any circuit or component that uses semiconductors or semiconductor technology for operation.

[2] An electrically operated device that mechanically switches electric circuits.

[3] A group of adjacent bits usually operated on as a unit, such as when moving data to and from memory. There are eight bits to a byte.

[4] A magnetic device used to convert electrical energy into linear motion.

[5] A bidirectional thyristor device used to control AC voltage.

[6] A device that uses the displacement of air in a bellows or diaphragm to produce time delay.

Acknowledgments

There are four volumes and three manuals in the *Wordworks™ Series*. In order to properly acknowledge the many people and organizations that have contributed to this project, I have chosen to first extend my thanks to those who assisted the endeavor in a larger context. A great many more contributors helped shape the separate volumes.

The Wordworks™ Series

The *Wordworks™* project would never have come about without the patience and generosity of my colleague and assistant Patricia Britz. There were 4000 pages of manuscript, endless keyboarding tasks, and elaborate page spreads. The project would have been impossible without Pat.

My colleague, Dr. Rita Smilkstein (published by Harcourt Brace), deserves very special thanks. Dr. Smilkstein has read every manuscript page of the project and contributed endless ideas and support.

To Stephen Helba, editor-in-chief at Prentice Hall, Pearson Education, I owe a special thanks. Some years ago I came home one day and found twenty-five pages of contracts spread all over the floor under an old fax machine. From the beginning Stephen took a personal interest in the *Wordworks™* project and has been a source of encouragement for the many years it has taken to develop.

To Dr. David Mitchell, President of South Seattle Community College, I owe thanks for a special favor. When he was the former dean of my campus, I sought his help in finding a space where I could create a dedicated classroom for my program. His response was immediate. We surveyed a prospect and agreed to the experiment. He budgeted a fully provisioned room of tables and chairs and blackboards and other features I requested. That dedicated teaching and learning environment played a major role in developing the concepts embodied in the *Wordworks™* series.

I would like to thank Marc Vassallo, Lee Anne White, Jeff Kolle, and other editors of the Taunton Press in Newtown, Connecticut. It was their faith in my ability to produce cover stories, feature articles, and shorter pieces for the nationally known Taunton magazines that give me the gumption to try to write the *Wordworks™*. It was a great boost to realize that each publication had a circulation of several hundred thousand readers and that, all told,

I had reached several million readers thanks to the Taunton staff. I also discovered the excitement of working with publishing teams and writing copy that included photographic compositions and line art concepts.

I would like to thank a number of my colleagues for their technical advice: Steve Anderson (physics), Lynn Arnold (network technology), Dale Cook (HVAC), Fred Edelman (CAD), Tom Griffith (chemistry), John Hagans (computer technology), Ralph Jenne (mathematics), Chris Sanders and Dennis Schaffer (electronics).

It helps to have the support of a close friend when facing the misgivings involved in an enormous project. Rob Vinnedge, one of the Northwest's finest professional photographers, assisted me as I developed articles for the Taunton Press even when he was trying to meet his own book and magazine deadlines that were taking him as far away as Hong Kong. I regret that there was no time in our busy lives to share any shoots for *Wordworks,* but the deadlines were too tight. What matters the most is that Rob said, again, again, and again, "You can do it."

Writer's Handbook for Engineering Technicians and Technologists

My work on the handbook was greatly assisted by Dr. Rita Smilkstein, who graciously dedicated the time for *two* close readings of the complete text.

My special thanks to my friend Dr. Philip Klein, professor of linguistics at the University of Iowa, for examining sections of the text that I asked him to discuss with me.

A special note of gratitude to Corin Carper, a recent graduate in computer engineering, who checked the 180 web sites identified in the text. Although the sites come and go, Corin did his best to provide a timely update.

This text does not include the student samples that are a characteristic feature of *Wordworks,* but it does contain passages from the work of one former student. Ben Minson kindly permitted me to use the excerpts from one of his projects.

There are additional sources of material I would like to acknowledge.

Corporations and Publishers

Special thanks to the following corporations for permission to reproduce images of their products:

Alaris Medical System

Microsoft Corporation

Web Sites

I would like to thank the following for permission to use images from their web sites:

Building Site Network

The Heath Network/Galaxy

University of Saskatchewan

WWW Virtual Library

Yahoo! Inc.

Societies

The following societies and associations gave me permission to use images from their web sites:

Acoustic Society of America (member of the American Institute of Physics)

American Institute of Chemical Engineers

American National Standards Institute

American Society of Heating, Refrigerating, and Air-Conditioning Engineers

Association for Computing Machinery

Construction Specifications Institute

Institute of Electrical and Electronics Engineers

Two additional societies gave me permission to use calculations from their publications:

American Mathematical Society

American Society of Agricultural Engineers

Prentice Hall/Pearson Education

Special thanks to the production staff at Prentice Hall in Columbus, Ohio, and the Clarinda Company of St. Paul, Minnesota. Two production teams saw to it that the *Wordworks Series* would go to press. At Prentice Hall, I would like to acknowledge the following staff:

Editor in Chief: Stephen Helba

Associate Editor: Michelle Churma

Production Editor: Louise N. Sette

Design Coordinator: Robin G. Chukes

Cover Designer: Ceri Fitzgerald

Production Manager: Brian Fox

Marketing Manager: Jimmy Stephens

At Clarinda I would like to thank additional personnel for seeing the *Wordworks*™ series through the production stages:

Manager of Publication Services: Cindy Miller

Account Manager: Jennifer Graham

For her patience and skills, a special thanks to Barabara Liguori for copyediting the *Wordworks*™ series.

I would also like to acknowledge the reviewers of this text:

Pemela S. Ecker, Cincinnati State Technical and Community College (OH)

Dr. Harold P. Erickson, Lake Superior College (MN)

Patricia Evenson, Northcentral Technical College (WI)

Anne Gervasi, North Lake College (TX)

Mary Francis Gibbons, Richland College (TX)

Charles F. Kemnitz, Penn College of Technology (PA)

Diane Minger, Cedar Valley College (TX)

Gerald Nix, San Juan College (NM)

M. Craig Sanders, Bellevue Community College (WA)

Laurie Shapiro, Miami-Dade Community College (FL)

Richard L. Steil, Southwest School of Electronics (TX)

David K. Vaughan, Air Force Institute of Technology (OH)

Dave W. Rigby

Index

Editing Code

abb	abbreviation *(Chapter 7)*		**mm**	misplaced modifier *(Chapter 6)*
acc	check accuracy *(context error)*		**ms**	manuscript form *(layout error)* *See Basic Composition Skills, Chapter 7* *See Technical Document Basics, Chapter 2*
ad	adjective/adverb *Adjectives (Chapter 4)* *Adverbs (Chapter 4)*			
agr	agreement *(Chapter 6)*		**num**	number *(Chapter 7)*
awk	awkward *(context error)*		**ok**	let stand *(correct as is)*
bw	better word available *(context error)*		**¶**	paragraph *(context error)* *See Basic Composition Skills, Chapter 10*
cap	capitalization *(Chapter 5)*			
co	coordination *(Chapter 2)*		**no ¶**	no paragraph *(context error)* *See Basic Composition Skills, Chapter 10*
coh	coherence *(context problem)*			
conn	connotation *(Chapter 2)*		**//**	faulty parallelism *(Chapter 2)*
conc	conclusion *See Basic Composition Skills, Chapter 10** *See Technical Document Basics, Chapter 2*		**para**	paraphrase *(Chapter 6)*
			pass	passive construction *(Chapter 6)*
cs	comma splice *(Chapter 6)*		**ref**	unclear pronoun reference *(Chapter 6)*
def	define *(context clarification)*		**run-on**	run-on or fused sentence *(Chapter 6)*

*See the suggested chapters in Basic Composition Skills for Engineering Technicians and Technologists, another volume in the wordworks series.

dev	develop text *(context problem)*	**shift**	shift *person (Chapter 2)* *tense (Chapter 6)*
dm	dangling modifier *(Chapter 6)*	**slang**	slang *(Chapter 2)*
doc	documentation needed *(context problem)*	**slant**	language bias *(Chapter 2)*
ex?	example needed *(context problem)*	**sp**	spelling *(Chapter 5)*
frag	sentence fragment *(Chapter 6)*	**sub**	subordination *(Chapter 2)*
fs	fused sentence *(Chapter 6)*	**sum**	summarize *(context suggestion)*
id	idiom *(Chapter 2)*	**tr**	transpose *(context error)*
inc	incomplete construction *(context error)*	**trans**	transition *(context error)* *See Basic Composition* *Skills, Chapter 10*
intro	introduction *(context problem)* *See Basic Composition* *Skills, Chapter 10*	**vague**	vague *(context error)*
ital	italics *(Chapter 3)*	**verb**	verb form *(Chapter 4)*
jarg	jargon *(Chapter 2)*	**wrdy**	wordy *(style error)*
lc	lowercase (no cap) *(Chapter 5)*	**ww**	wrong word *(context error)*
log	logic *(possible context error)*	**ww?**	check word use *(context error)*